Fiddler on the Hoof

Fiddler on the Hoof

Around Ireland on Horseback with a Violin

Cathleen Leonard

Fiddler on the Hoof
Around Ireland on Horseback with a Violin

Copyright © 2020 Cathleen Leonard

All rights reserved. No part of this book may be reproduced or used in any form without the express written permission of the author. The only exception is by a reviewer, who may quote short excerpts in a published review.

Cover Image: The Road from Mizen Head, photo by the author
Cover Design created by the author using Kindle Cover Creator

The maps used in this book are licensed under Creative Commissions Attribution 3.0, and have been edited to show the route by Adriaan Kastelein. Maps originally created by Equestenebrarum.

Editing and proofreading by Sari Maydew

ISBN Paperback: 9798614000134

A Strange Request
www.AStrangeRequest.co.uk

www.facebook.com/AStrangeRequest

Printed by Amazon

'Faeries, come take me out of this dull world,
For I would ride with you upon the wind,
Run on top of the dishevelled tide,
And dance upon the mountains like a flame.'
– William Butler Yeats, The Land of Heart's Desire

Dedicated to the memory of my father, Tom Leonard, who first introduced me to the Irish myths and legends, and to traditional Irish music.

Contents

Acknowledgements..ix

Part 1: Cornwall to Wales

1: Reasons..1
2: Vlad...7
3: The Best Laid Plans...12
4: The £100 Horse...15
5: Setting Off..21
6: Meldon Viaduct..26
7: The West Country..30
8: The Somerset Levels......................................34
9: Settling In..38
10: Lumps and Bumps..41
11: Tracks and Trails..45
12: The Forest of Dean.......................................51
13: New Shoes for Dakota..................................57
14: The Black Mountains...................................62
15: The Brecon Beacons....................................68
16: A Mishap in the Mountains.........................72
17: Anything is Better than a Tent....................76
18: A Chance Encounter...................................80
19: Plas Dwbl..84

Part 2: Ireland

20: First Impressions of Ireland..93
21: The Baltimore Regatta...96
22: Mizen Head...101
23: The Cork and Kerry Mountains.................................107
24: The Black Valley and the Gap of Dunloe.....................113
25: Dingle..117
26: Travellers in Kerry..124
27: Mícheál..131
28: East Clare..139
29: Back on the Road..145
30: Mícheál's First Days on the Road..............................149
31: Galway..154
32: Roscommon...160
33: Leitrim...166
34: An Afternoon with the Lennons................................171
35: Borders...176
36: Donegal...181
37: Finish Lines..188
38: Conundrums..194
39: Farewell to Ireland...203
Epilogue

Acknowledgements

There are so many people who helped us, both on the journey and in the writing of this book that I hardly know where to begin. Without all their support, encouragement, kindness and hospitality this journey would never have happened, and neither would this book.

Firstly, I'd like to thank all my friends and family, who patiently tolerate all my fanciful ideas and even go so far as to encourage me – both in my writing, and in the pursuit of my dreams. You are too numerous to list here, but your efforts and contributions do not go unnoticed.

To CuChullaine O'Reilly of the Long Riders' Guild: my deepest gratitude for your unwavering support throughout and as ever, for providing such a wonderful source of inspiration for dreamers and adventurers such as myself.

A huge thank you to the many lovely people who hosted us along the way: your kindness, generosity, and hospitality kept our spirits high, and bolstered our faith in humanity throughout the journey. Roughly in order of appearance, our thanks goes to: Caro Woods; Liz Parks and Guy Cashmore; Emma and Andrew Bowyer; Lisa Walker; The Conquest Centre; April Anderson; Cheryl Green; Guy Clifford; Mel and Woody; Anna Seymour; Maggie Brombley; Hazel and Pete Weyman; Mike Cromie; Lisanne and all at the Red Horse Foundation; Carolyn and Geoff; Katherine and David Nelson-Brown, and all at Perrygrove Railway; Morgan; everyone at Triley Fields Equestrian Centre; Marie near Grwyne Fawr Reservoir; the Price family; Sarah and Glanville of the Dan Yr Ogof Show Caves and Shire Horse Centre; Jaqui and Phil; Vic and Roger; the man near Llanpumsaint – I hope you found happiness, not just a housekeeper!; Alan and Judith Sterling – we are forever indebted to you for your help; Laura and all at Plas Dwbl; Nathan Deakin Transport; West Cork Equine Centre; Tullineaskey Equestrian Centre; Jane Scully; Linda and Jimmy O'Donovan; Dennis and Mary; the friendly staff in the visitor's centre at Mizen Head; the nice family near Durrus; Tim and Sandra; all the friendly people of Kilgarvan; Gene in the Black Valley; the jarveys at Kate Kearney's Cottage; Eric; Mamood; everyone at Dingle Camphill Community; Serge and Lisa; Louis and Sandra; Kevin and Kristiana; the couple near Crecora; Liz and Joe Ryan; my mother, Roxanne Leonard; Nard, Tonia, and Mieka for hosting the horses in Mountshannon; Agnes and Mike; the nice family near Peterswell; Suzanne and Matthew; farmer Paul; farmer James near Skehana; Kevin Higgin and Geraldine at Mountbellew; Margaret and Brendan; Broc and Edel; the farmer near Elphin; Gerry and Mary Gilhooly; the nice family near Lough Allen; John and Joy; Declan Sweeney; Paul and Catherine Keogh; Wendy and Francie; farmers near Killygordon; Patricia, Robert, and Nicola near Raphoe; William near Sappagh; James near Dumfries; Derek McLaughlin for letting the horses stay so long; and finally a massive thank you to Toni and Jack Harkin for putting up with us moping about your house in that last difficult week. Also a huge thank you to Adam Firks for taking a chance and helping two random strangers get home from a mad adventure!

Thanks to all the farriers who shod our horses throughout the journey: neither of the horses were easy and your patience and gentleness was appreciated throughout.

Thank you to all the people who shared the myths, legends, local history, folktales, and stories of their own that lent so much colour to our journey and helped bring Ireland's magic alive. To all the musicians we met along the way who shared tunes with us, thank you; and a huge thank you especially to Ben and Maurice Lennon for playing me so many beautiful melodies, for improving my bust-up old fiddle, and for giving me so much encouragement and kind feedback on my playing. My musical career is still on hold while my equestrian wanderlust runs its course, but I will one day settle down and give it my all. Thank you also to my mother for making me learn the violin in the first place. You were right: it is indeed a good skill to have.

In the writing of this book I would like to say thank you to Sari Maydew for all her hard work editing the manuscript, and thank you to friends and family who read through the book in its early stages and provided some much-welcomed feedback. Thank you also to my aunt, Linda Williamson, for proofreading the final manuscript and picking up on the little things that were missed in the editing process.

To Adriaan Kastelein: thank you for creating such beautiful maps for the book.

I should probably also thank Sarah and Pedro Krämer, who have hosted us in Portugal for the winter and who have put up with me writing for hours and hours on end when I should probably have been pulling my weight a bit more about the farm! Thank you for your patience and understanding.

Finally, and most importantly, I would like to thank my companions Vlad, Spirit, Oisín, Dakota, and little Micheál. I am indebted to you all for patiently putting up with my madness, and for your absolute willingness to follow me against all the odds in order to make my wildest dreams come true.

Preface

Ireland is a country rich in folklore, fairy tales, legends, and music. It abounds with a unique enchantment that has captured the imaginations of many, and has always held a strong appeal for me.

In the following pages, I have tried to give the reader a taste of Ireland's magic, as well as providing an account of our own journey to, and across, the Emerald Isle. There are hundreds of tales I would have liked to tell about all the places we went through – but I could not fit them all into this book. Instead I have included snippets here and there of the stories most relevant to our journey. Although I have barely scratched the surface, and have certainly not done the tales as much justice as they deserved, I hope perhaps the reader will feel inspired to go in search of more of the Celtic myths, legends and fairy tales, and thus discover for themselves some of Ireland's magic.

Part 1: Cornwall to Wales

The route from Cornwall to Wales

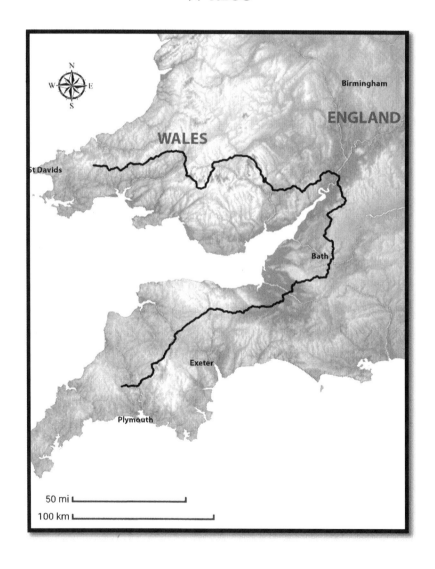

1

Reasons

The* wind was stripping dead leaves off the trees and blowing them in swirling gusts across the asphalt where a handful of familiar cars were parked. Taking a deep breath, I got out of my van and headed for the pub door. I heard the familiar click of the latch as I pushed the door open and stepped inside. It was dark in the dingy bar, a fire roared in the wood-burning stove, and brass ornaments and harness buckles glinted on the dark wooden beams. Around the tall wooden island in the middle of the bar, the usual crowd of regulars sat perched on the same stools they always occupied, drinking pints of the same beer they always drank, having the same old conversations they always had – every day of the week.

It was the twenty-eighth of October, the day after I'd returned from my nine week long, 1,000 mile solo journey from Scotland to Cornwall with my steadfast old draught horse, Taliesin. It had been the adventure of a lifetime and a childhood dream finally realised.

I'd returned quietly and reluctantly, dragging my feet and wishing I were headed anywhere but home. I hadn't wanted the adventure to end. There's something about life on the road – the simplicity and the

---

* For photos please visit www.AStrangeRequest.co.uk

freedom of it, coupled with the insecurity of never knowing what to expect – that makes you feel so intensely alive. But winter and bad weather had been close on our heels and Taliesin had needed a rest.

Before leaving for Scotland, I'd been working behind the bar of that little village pub. It was a typical pub in rural Cornwall in as much as it was full of staunch regulars by whose drinking habits you could set a clock. It was the sort of place where everyone knew everyone else's business and what they didn't know, they would make up. Actually, it had been long hours of pulling pints and listening to mindless gossip that had finally driven me to chase my dream of riding the length of the country. I'd been afraid that if I didn't do it I would become stuck – sucked into a rut of mundane trivia and silly village drama while life passed me by. In the end, that idea had been more terrifying than the prospect of riding alone from one end of the country to the other.

The day after my return, I'd gone to see about getting my job back; but the moment I walked through the pub door and saw the same old bunch of people, doing the same old things they always did, and having the same old conversations they always had, I knew I couldn't face it again. It made me feel as though I had never been away, as though the journey had never happened. Nothing here had changed. Not a single thing. Except me. No, I decided firmly as I turned around and headed for the door, I needed to do something more meaningful and more rewarding with my time.

A few weeks later I took up employment in the only other sector in Cornwall where there is never a shortage of work: the healthcare industry.

Long days of driving from the home of one client to the next ensued; tending to the frail and elderly, the disabled, the sick, and the dying; helping them to perform the basic daily tasks which I – in my relatively healthy twenty-seven-year-old body – naively took for granted. It was hard work, with early morning starts and late night finishes, and the pay left a lot to be desired. Although the work was far from glamorous, I enjoyed making a difference – however small – to those people's lives. I loved hearing their stories, too – the tales of their youth and life in the 'good old days'. I happily listened to the life-lessons and hard-won wisdom that they offered up in our daily encounters while I assisted them to wash and dress, prepared meals, and served endless cups of tea.

The job broadened my own perspective on life and opened my eyes to the harsh realities of sickness and old age. It taught me the valuable lesson that health and mobility should be cherished above all other things, and that neither should ever be taken for granted.

Something else happened not long after my return which further hammered home the stark lesson that nothing in life should be taken for granted, and that was my father's illness.

He had known for a while that something was wrong – he'd said as much over the phone on the few occasions I'd spoken to him on my long journey from Scotland to Cornwall. In the last week of the adventure I rang him as I was settling in for the night at a stop near Taunton. He told me he was having trouble eating and was in a lot of discomfort when he did.

'I'm worried,' he said, his voice sounding thin and weak over the phone. 'What will your mother do if something happens to me?'

Always the pessimist! I thought, rolling my eyes.

'Don't be like that, Dad. I'm sure it's nothing to worry about. You'll be fine!' I said almost flippantly.

He and mum were living in County Clare in the west of Ireland, and things over there happen slowly, so it wasn't until some time in the middle of November that he managed to get an appointment to find out what was causing the problem. More than a month passed while we waited for the results, which finally came through just days before Christmas. It wasn't something straightforward at all – it was stomach cancer.

The news was sudden and unexpected. I suppose you never really believe something like that will happen to you or your immediate loved ones. It's one of those things that happen to other people, and other families.

Yet here it was: cancer.

Another long wait for scans and an appointment with a specialist followed. It was Christmas and everything in Ireland grinds to a halt. There was nothing available until the New Year. By the time Dad finally got to see a consultant in early January 2018, it was only to be told that the cancer had spread too far to be operable. It was untreatable so they sent him home to die.

    Dad wasn't going to give up that easily, though. Not without a fight first. There were alternative treatments out there – things the medical profession wouldn't acknowledge. What was the harm in trying? And so began a daily cocktail of colloidal silver, hemp oil, CBD oil, sodium bicarbonate, and a whole pile of other alternative and herbal remedies whose names and purposes I have now forgotten – all accompanied by regular healing sessions. But to no avail.

    Barely able to eat, Dad gradually wasted away to nothing. He looked like a skeleton, but for fluid that was accumulating around his stomach. It needed to be drained. It was a routine procedure to make him more comfortable, they said, and they could do it at Limerick Hospice. He would be out in a day or two.

    A day or two became a week, and then two. He was growing frailer by the day and was too weak to go home, the doctors said. Besides, they could care for him better in the hospice. He was starting to fade now, drifting further into his own space and time, far away from the real world; he was awake but not entirely present. I'd seen this in my work with people nearing the end of life. Our conversations over the phone gradually became more rambling and less comprehensible. It was hard to follow what Dad was saying as he wandered listlessly from one unrelated topic to the next. Every now and then he would become lucid and talk coherently for a while, before drifting off again to who-knew-where – and each time we spoke his voice sounded thinner and feebler.

    Dad never left the hospice and, on the 12th of February 2018, he passed away. He was only sixty-nine years old. I was with him when he died, sitting by his bed holding his hand for the last few hours of his life, listening to each laboured breath as he struggled to escape his cumbersome, cancer-riddled body. It's a hard fight to let go at the end, and I was glad to be there, surrounded by his friends and family. It was a peaceful and painless passing, and for that I was grateful. Working with people who had spent years suffering with degenerative, and utterly debilitating illnesses had shown me all too clearly what a slow and painful death might have been – both for him, and for the rest of my family. Sometimes you have to be grateful for the small mercies.

If care work had taught me the value of health and mobility, then losing my father taught me the value of time. It reinforced my already strong determination to live life to the fullest and to dedicate whatever time I might have to the pursuit of my dreams. My personal motto: 'Do it while you can!' took on a whole new meaning.

So what was I waiting for? There was a whole world out there to explore! And the first stop, I decided, would be Ireland.

The little island on the western-most edge of Europe, with its many long-fingered peninsulas reaching into the expanse of the wild Atlantic, had always held a certain appeal for me.

Ireland was a land steeped in the myths and legends of the Celtic race, where ancient gods and heroes of times long past had played out their epic sagas. It was a place where folklore and fairy tales abounded and where people still held a strong belief in the fairy folk who wandered the hollow hills and danced around lone thorn-trees on moonlit nights; and where tinkers in horse-drawn wagons spun stories around turf fires on the roadsides.

That was a rather romantic image, so you can probably imagine how disappointed I was when, on visiting Ireland and having a good look around after my parents moved there in 2012, I found it to be … well, pretty much like any other country I'd been to. There were no wild men charging about on horseback doing battle with gods and giants, or undertaking heroic quests; no fairies frolicking about in the bushes at night, and hardly a trace of anyone wandering the mountain roads in horse-drawn wagons. Instead there were just the usual kinds of people living pretty normal lives. Yet even in the tamed and settled landscape the legends and myths lived on – immortalised in the names of the hills, the lakes, the rivers, and the sea-ravaged headlands, their ancient memory echoing down the ages.

And then there was the music! Music that held the story of the land and the people who had lived there; beautiful, evocative music! Captivating melodies that bore the names of people and places from all over Ireland, or ballads that told of the hardships and heartbreaks that had befallen men and women who were long-since dead but whose stories were kept alive through songs that had spread to the four corners of the world.

My father was Irish and I grew up listening to the music of the Dubliners, Planxty, the Chieftains, and many, many more of the artists that had helped to popularise traditional Irish music in the latter part of the 20th century. I loved the songs, and even more than that, I loved the instrumental melodies: the jigs and reels, polkas and hornpipes, airs and marches. Some say that many of the most beautiful Irish tunes were actually fairy melodies, picked up by eavesdropping mortals. I could well believe it! It was the kind of music that never failed to set your feet tapping, and brought a smile to your face. It had provided a soundtrack to my childish imaginings in which I roamed the wild hills on horseback with a wolf at my side, setting off on all kinds of epic adventures.

I wanted to discover the country that had given rise to such beautiful music and so many magical tales; I wanted to meet its people and see for myself some of the mythical places of folklore and legend; and above all – as a fiddle player – I wanted to map a musical trail across that enchanted island, discovering the tunes of the places through which we passed. What better way to do this than on horseback, immersing myself in the landscape, and putting myself at the mercy of its people?

It was time to hit the road again, and this adventure, I decided, would be a musical journey on horseback and a quest of my own to discover ancient Ireland.

2

Vlad

I hadn't been back long from my Scotland to Cornwall adventure when, sitting in a friend's kitchen one day drinking coffee and putting the world to rights, she announced that she'd let her spare room to a young Romanian man who was a lecturer at the local agricultural college.

My ears pricked.

One of my many dreams was to travel the length of the Carpathian Mountains on horseback, to experience a wilderness where bears, wolves, and lynx roamed freely – and I wanted to start that ride in Romania. That was about as far as I'd got with my plans because I'd decided to ride from Scotland to Cornwall instead. It had seemed a comparatively more sensible idea as I knew nothing about Eastern Europe, didn't speak any of the relevant languages and had no idea how to go about finding horses, equipment, or even how to plan a route there. What were the roads like? What access laws were there for the countryside? Could you ride and camp anywhere? What were the people like? How would I be received?

What I really needed were contacts – some local knowledge, a bit of inside information, and some help learning the language. By a pleasant stroke of luck the opportunity to gain all of the above was being dropped right into my lap! And as it happened Vlad had already heard of me:

'I was driving home from work one evening and, looking for a distraction from my thoughts, I tuned into the local radio station. A young woman was giving an interview about a journey she'd just completed, riding her horse from Scotland to Cornwall. Who in their right mind would want to do such a thing, I wondered? And why solo? That wouldn't be much fun, I thought, recalling the kayaking expeditions I'd done with my friends along the Danube River.

'Definitely mad, I decided, as she described her adventures. Yet her answers were coherent and to the point. There was some mention of charities, too. She sounded like a strong person, and fiercely determined. Perhaps not so crazy after all...! The radio signal wavered and the voice in the speakers faded away as I drove across Bodmin Moor, so I made a mental note to go back and listen to the interview again online once I got home.

'I had just taken a room in an old barn conversion belonging to a retired Dutch vet. It was only a temporary stop while I figured out my next move. I was at a point in my life where my teaching career was bringing me more stress than satisfaction due to the continuing government cuts to the education system, and a failing relationship with my boss. I was ready to give up everything I'd worked for in the UK over the last decade and head back to Romania. I wanted to buy some land where I could live a simpler, more natural life – but I needed to tie up the loose ends first.

'A few days went by, and I forgot about the interview until one afternoon, whilst washing the dishes, a young woman entered the kitchen.

'"This is my friend Cathleen," my landlady José said, introducing us. "She's just got back from riding her horse from Scotland to Cornwall."'

Vlad

Some weeks later Vlad left his job teaching and, like me, found himself at a bit of a loose end so we began spending time together – discussing adventures while I picked his brains about Romania. We quickly discovered some common dreams and interests and it wasn't long before one thing led to another and we became something of an item.

I was quite wary about getting involved with someone again. Past relationships had mostly been one emotional disaster after another that had left me rather sceptical of the whole ordeal. I was cautious lest I get tied down by someone who would ultimately try to restrict my freedom, strip me of my autonomy, sap my energy, and require far too much of that dreaded word: compromise – only for the whole thing to end in a quivering pile of emotional trauma that I'd then have to spend the next few years recovering from. No thank you! I had walked blindly into that one before and I wasn't keen to put myself through it all again. And besides, where would I find someone crazy enough to want to travel the world on horseback with me?

For all my reluctance and lack of enthusiasm, Vlad was undeterred. He persevered, bent over backwards, and would have offered me the moon on a stick if that would have convinced me of the depth and sincerity of his feelings. Endless bunches of flowers, romantic dinners, grand promises, and frequent declarations of his undying love only fell on deaf ears or were met with eye-rolls and scepticism. What actually did impress me about Vlad, however, was his willingness to wade through a thick soup of mud in the driving rain pushing barrow-loads of horse-muck; or his readiness to fill hay-nets on cold, dark, winter nights. He would frequently turn up at the stables at unsociable hours when I was tired and hungry at the end of a long shift at work to help me feed and muck out the horses, and then offer me a lovingly prepared hot meal … but even that wasn't enough to convince me about the whole relationship deal.

Throughout my father's illness Vlad offered support in every way he could, driving me to and from the airport and looking after the horses while I spent time with Dad in his final days and then stayed on to help Mum with the funeral arrangements. Still I did everything I could to keep Vlad at arm's length, expecting it all to fall apart at any given moment. Poor old Vlad! It is a real testament to his perseverance, determination, and strength of character that he stuck unwaveringly by me through all those difficult months. No matter what I threw at him, or how hard I tried to push him away, he'd simply bounce back with renewed enthusiasm.

When I told him of my plan to ride around Ireland that summer he hinted that he would like to come along.

'Join me for a week or two,' I offered. 'You'll soon know whether you like it.'

'Let me sleep on it,' Vlad said, suddenly uncertain.

By the next morning he had made up his mind.

'I'm coming with you!' he announced with conviction.

There was only one glitch: Vlad had never ridden a horse before!

Lessons began in earnest. We started with Taliesin – he being my most solid, dependable horse. Taliesin was generally of the opinion that doing anything more than a slow, steady walk was a total waste of energy. He was safe, sensible, and up to the job, I thought.

Taliesin tolerated his new role as 'School Horse' for all of about three rides and then decided it wasn't for him. He expressed this sentiment by throwing in some uncharacteristic bucks. At first I thought his back might be hurting, but when I got on to ride him he was fine. Having failed with the bucking, Taliesin then took to swinging his head round and biting Vlad's legs whenever I wasn't looking; and when that didn't work he resorted to snatching the reins from Vlad's hands, putting his head down to graze, and staunchly refusing to budge unless it was towards home. Clearly this wasn't going to work.

In the time that Vlad had been helping with the horses, he'd started to develop a bond with Oisín (Usheen), my other French draught gelding, who – although of a similar size and breed to Taliesin – had a completely different temperament. Vlad liked Oisín's cheeky character and over-enthusiastic approach to life, and he felt they might get on well together. I wasn't so convinced. Oisín could be rather excitable and liked to break into short, energetic bursts of trot at random intervals, with the odd buck thrown in here and there for good measure when he was feeling especially pleased with himself. None of it was done with any malice, I must add, but it wasn't ideal for a novice rider. However, since Taliesin wasn't co-operating I agreed to give it a go. To my surprise, Oisín kindly obliged and generally refrained from bucking or bouncing too much when Vlad was in the saddle, and thus Vlad's riding slowly improved.

In this way, to the great relief of my friends and family, I gained a companion; at least for the first part of my journey.

3

The Best Laid Plans

I'm not a great lover of plans – not rigid ones anyway – because you can almost guarantee that the minute you make one, something will come along to scupper it. That said, however, it is necessary to have a vague idea of what you're about to do, if only so you don't sound like a complete air-headed dreamer.

The plan we made went something like this:

- Where? Ireland.

- When? The end of June – because in my experience the start of June is often wet, and if there's one thing I hate it's sleeping in a soggy tent full of soggy gear.

- How? Ride the horses to Wales to save money on transport, and worry about the logistics of getting over the sea to Ireland when we got nearer to Pembroke. Once in Ireland, we'd ride from the southernmost point to the northernmost point, just to say we'd done the lot.

- Who? Taliesin and Oisín – my two ten-year-old French draught horses whom I'd rescued from slaughter when they were foals. They were both experienced travel horses, both were sane, sensible, and bombproof. I trusted them implicitly. Vlad would ride with me as far as Pembroke and decide from there whether he wanted to continue on to Ireland.

- Why? That old biscuit! Why does anyone need a reason to head off on a mad adventure other than just for the sheer hell of it? Well apparently people want reasons. In my experience it's easier to support a charitable cause because people seem to understand the journey better if you do. Besides, if you're going to do something crazy and someone somewhere can benefit from it, then why not? We decided to use this journey to help out two of the equine assisted therapy charities that had so kindly put me and Taliesin up on our way home from Scotland the year before. One was the Red Horse Foundation in Stroud, and the other was the Conquest Centre near Taunton.

'What about raising money for the hospice where Dad died?' my sister asked. 'They rely on donations. It would be nice to give something back.'

I approached the hospice and told them the plan. At first they were all for it, but then a week or so later I received a phone-call from the fundraising co-ordinator.

'There's been an unfortunate incident involving a llama and some undesirables here in Limerick,' she said in a lilting West of Ireland accent.

Incident involving a llama? Undesirables? A series of bizarre and rather comical images raced through my head and I fought hard to suppress a giggle. Clearly this was no laughing matter.

'Oh dear!' I said, trying to sound appropriately concerned.

'I'm afraid on this occasion we won't be able to give you the go-ahead for your fundraising idea until it all blows over,' she said, without elaborating on the llama situation.

So much for that idea then!

The Conquest Centre and the Red Horse Foundation had no such objections, and thus they became the sole beneficiaries of our fundraising endeavours.

So that was the plan in all its vague and undefined glory. As for the rest? Leave it up to Fate!

They say pride comes before a fall. I think the same can probably be said for confidence.

I wasn't worried about setting off this time, and I wasn't nervous. As a seasoned equestrian traveller I knew what the trip entailed. I knew what to expect from the road, and understood the challenges that lay ahead both for me and for the horses. I knew what equipment to take, knew my animals and how they coped, knew myself and how I coped; I even knew the first two hundred or so miles of the route and most of the people we would be staying with. It is safe to say I was confident that unless something went seriously wrong, we would be just fine. There was no chance of falling at the first hurdle, or turning back at the first sign of trouble – I was tougher than that now, and the animals were too. Even Vlad, I felt, was made of sterner stuff than I'd ever given him credit for – he'd proven his strength and determination time and again over the months that I'd known him. We were unquestionably ready for this.

That is until Taliesin went suddenly, and inexplicably, lame.

4

The £100 Horse

It was two days before we were due to leave, the farrier was booked for the following day to shoe the horses, and we were on our final training ride. After the first mile I noticed that, although fine walking and trotting in a straight line, Taliesin was hopping lame on his left hind leg when turning corners. This was the last thing we needed two days before setting off on a long distance journey!

'He's probably just tweaked something galloping about in the field. The ground's so hard at the moment,' my horsey friends tried to reassure me. It didn't work. We couldn't set off on a long ride with a horse that wasn't a hundred per cent sound – that would be irresponsible and cruel. We couldn't take Oisín on his own, either. That had been a disaster two years previously when I'd ridden him to Land's End. Should we postpone leaving and hope that Taliesin made a speedy recovery? What if he didn't? And if he did come sound, what if he went lame again as soon as we hit the road? We couldn't risk it.

'Well, there's always Dakota!' I said jokingly to Vlad when we got back to the yard.

I was referring to my seven-year-old skewbald gelding, a highly-strung bundle of nerves who thought the world was out to get him. I'd broken my collarbone throwing myself off him at a flat out gallop two years before, following a minor disagreement about speed and direction. It had shattered my confidence in him and although I'd ridden him a couple of times since, he was still very green and incredibly nervous – especially of people! Not exactly the dependable, level-headed kind of horse you want for a long journey into the unknown!

I hoped desperately that Taliesin would at least show some signs of improvement by the morning when the farrier came to give his professional opinion on the situation.

He didn't. If anything, he was worse.

'It's definitely not in his hoof,' the farrier announced after a thorough examination.

Damn! I'd been hoping for something straightforward like an abscess. Now what? I hadn't actually been serious about taking Dakota. It was a ludicrous idea at best, and downright dangerous at worst!

Sired by my friend Jan's stallion – a coloured warmblood from royal bloodlines – and out of a Welsh Section C driving mare that belonged to an elderly farmer named Mitch, Dakota had spent the first three years of his life running feral with a herd of mares and foals. In those years he'd had very little to do with people at all. We would go over regularly to check on the herd and to feed them in the winter, but while the other mares would come up to us for food and fuss, Dakota's dam would head off to the far side of the field taking her foals with her. Somehow through her behaviour all her foals developed a deep-rooted aversion to people.

One day, in the late autumn of 2013, I'd gone over to Mitch's farm where the mares and foals were kept to help Jan round them all up and separate out the colts that needed handling and gelding ready to be sold. I looked over at a chestnut and white skewbald colt as he trotted gracefully around the open barn and was struck by how pretty he was and how gracefully he moved.

'You're gorgeous!' I said to him. 'Do you want to come home with me?'

He threw his head up and down in a vigorous nod.

I laughed and asked him again – and again he nodded. It must be a sign!

The £100 Horse

'You can have him!' Mitch said from where he was leaning on the gate, crutch in hand. Mitch was in his eighties and not very steady on his legs. His days of handling youngstock were long gone. 'He's for sale.'

'Yeah, I'll give you fifty quid for him,' I said, winking at Jan. We both knew Mitch would want silly money for the horse. And anyway, I certainly didn't need another one!

'I'll take a hundred for him,' Mitch replied. I think Jan nearly fainted.

'I'll think about it,' I said.

When we got him back to Jan's yard, herded him off the lorry and into a stable, I looked at Dakota doubtfully. He wasn't really my type of horse. I like chunky, sensible cobs with a nice steady temperament, and an easy-going nature. This horse was fearful and wild, with the whites of his eyes permanently on show.

Jan wasn't all that good on her legs either – she'd broken too many bones in all her years working with polo ponies, racehorses, and breaking horses for a living, and she was getting on in years now. She'd not long had a knee replacement which was taking forever to heal, and she needed crutches to walk, so she asked me to handle the young horses for her. I loved working with the youngsters and getting them quiet, teaching them to accept a headcollar and to lead nicely, and allowing me to pick up their feet. It was very rewarding work.

In the weeks that followed, the other colts we'd brought home were handled, halter broken, and castrated – but Dakota remained tense and terrified, standing at the back of the stable warily eyeing up whoever was around and panicking madly whenever anyone came too close to him. Definitely not my kind of horse! I was glad I hadn't agreed to buy him, but with the other colts now calm and docile, there was only him left to tame. I'd procrastinated long enough, reluctant to make a connection with this little horse in case I became attached and wanted to buy him.

I put many hours into him, waiting until we'd finished all the feeding and mucking out and everyone had left. Then, in the stillness of the barn, where the only sounds were of horses happily chomping on hay and rustling about in their deep straw beds, I quietly, patiently, and oh-so-slowly began to make progress with this feral horse.

It started with holding a scoop of feed over the door of the stable and gently coaxing Dakota to eat from it. At first he would stand as far away as he could, craning his neck to snatch a quick mouthful of nuts before retreating quickly to the back of the stable again, but after a while his confidence grew and he would stand closer, more at ease. After a week or two, we progressed to me touching his nose while he ate and eventually I could stand inside the stable door with him.

The first time I touched his shoulder he leapt to the back of his box, snorting and quivering with terror as though I'd given him an electric shock, but when he'd recovered from the fright he slowly and cautiously came back to eat and I repeated the process until at last he accepted being stroked all down his neck and shoulder. Next I used a long stick to touch him all over his back, under his belly, and down his legs. He stood rigid, ready to kick out, just about bearing it until he finally realised I wasn't going to hurt him.

After months of quiet, steady work, I could run my hand down his neck and shoulders and all along his back while he ate. But I was the only person who could. No one else could get near him – his suspicion of people was just too deep-rooted.

'We can't sell him like this. No one will buy him,' Jan said one day as she watched me working with him. 'I think you'll have to have him.'

'I don't want him,' I said. What would I do with a horse like this? I wasn't capable or confident enough to work with this animal beyond getting him quiet and handled. I'd certainly never manage to ride him!

'I'll buy him for you!' Jan pushed. 'He'll have to be shot if you don't have him. It's the kindest thing to do.'

Dakota's two older siblings – a half brother and full sister – had been equally resistant to human contact. Both had ended up in the slaughterhouse because they'd been wild to the point of dangerous; I knew Jan meant what she said.

I have too soft a heart really, and after hours spent building up trust with Dakota, I had become quite fond of that quirky little horse – even if he was a lunatic! It would break my heart to see him shot. Under that tense and terrified exterior I felt sure there was a sweet and gentle personality waiting to come out, if only he could learn to relax enough around people. In the end I gave Mitch the fifty pounds I'd originally offered, Jan put up the rest of the money, and thus I acquired Dakota.

It took months to get a headcollar on him and getting him castrated had been little short of a nightmare. Dakota was terrified of the vet and had sent him flying across the stable with a kick to the shins when the vet tried to sedate him. He resisted three rounds of sedative and wouldn't go under, so in the end we had to tie him to a very sturdy pole with about four ropes while I held his front leg up to stop him kicking while the vet performed the operation. Once healed, it was time to bring Dakota home.

I introduced him slowly to my herd, one at a time. Oisín and Taliesin both took an instant dislike to the new member of the family. Taliesin even ran him through a fence by the scruff of the neck on their first meeting, but eventually they all calmed down and settled into herd life.

A year later, when he was four and a little more calm and relaxed, I began work to back him. He was still nervous and tense and wouldn't let anyone else near him, but he trusted me and I was amazed when he accepted everything I did to him without panicking – including letting me climb up on his back. He seemed to enjoy being ridden and going out exploring the quiet lanes and he was surprisingly sensible with traffic, too – although people were still seen as a threat and drain covers were given a wide berth, a protruding eye, and a loud, terrified snort.

Dakota had been going well for a few months when one day, whilst out riding him around the lanes in a bitless bridle, he started trotting and wouldn't stop. We were heading straight for a blind junction with a busy road and suddenly I had no control. This was definitely not going to end well!

Thinking fast, I managed to turn him into an open field, hoping that I might be able to get him under control and slow him down again. Instead, Dakota upped the gears and began to canter across the field. Reaching the far side, he turned around and headed back towards the road at a flat out gallop. I still had no control and we were now going faster than ever. With seconds to spare I decided to bail out and attempted to dismount. I hit the hard ground shoulder first, leapt up, and ran to the gate to stop Dakota getting out into the road. It wasn't until I reached the gateway that I noticed something was wrong. Should there be that scraping sensation in my collarbone when I move my left arm? And had there always been that lump sticking up under the skin? I nearly passed out.

After several weeks recovering from the injury, I was back in the saddle but I'd lost my nerve and barely rode Dakota in the years that followed. If I'm really honest, I was scared of him.

And that, dear readers, is the story of Dakota and should tell you everything you need to know about why the idea of taking him on a long ride for hundreds of miles through unknown territory was not an especially appealing one. In fact a less suitable horse for such a journey would be pretty hard to find! Yet the fact remained that with Taliesin out of the equation he was my only option – which would probably explain why I suddenly wasn't feeling all that confident any more!

5

Setting Off

The day of departure dawned bright and sunny. It was the twenty-seventh of June 2018, and Britain was baking in a heat wave. By eight o'clock the air was stiflingly hot and dry.

I spent the morning finishing off a few last-minute jobs and checking over our equipment one final time before I headed over to say goodbye to Taliesin, who was now facing a lazy summer in a field playing uncle to my friend's herd of young Dartmoor Hill Ponies. He seemed pretty pleased with his lot, and not in the least bit sorry for throwing such an enormous spanner in the works right at the last minute. I bid him a fond farewell and drove home with a heavy heart. I was going to miss him with his calm, steady nature, and easy-going, philosophical approach to existence. What on earth would life on the road be like with Dakota? And would either of us live to tell the tale? I tried not to think about it.

Vlad arrived on the yard at about 10 o'clock, his bulging saddle bags a great deal heavier than they'd been the night before when I'd carefully packed them with the bare minimum and most necessary of equipment.

'What have you got in these?' I asked incredulously.

'Only a few little things,' he shrugged.

'Vlad, this is ridiculous! We can't take all of this stuff,' I said, opening the packs and pulling out a kilo of dried onion, two tubes of Harissa paste, a tube of garlic paste, four litres of elderflower champagne, half a litre of elderflower cordial, and a bulky wash-bag packed full of unnecessary toiletries. I'd only just managed to convince him the night before that a disposable barbecue was a total impracticality that would definitely not fit in the saddlebags! Clearly our ideas of 'bare essentials' and 'travelling light' differed vastly.

'Poor Oisín! He'll collapse under the weight of it all!' I said, rolling my eyes at Vlad's protestations.

Never one to let anything go to waste, Vlad did his best to polish off all the elderflower champagne while we saddled the horses and loaded up the gear.

Packing was slow business. It was a first for Vlad – who had barely gotten to grips with putting a saddle on a horse, never mind packs – and it was a first for Dakota, who received the saddlebags with an expected amount of terror. He flapped wildly about at the end of his lead rope, with his eyes popping out on stalks, snorting loudly. Once he'd calmed down enough to realise the packs weren't actually a threat, he accepted his load, and I worked fast to get everything fixed firmly to the saddle just in case he suddenly changed his mind.

After the best part of three hours the horses were loaded up with everything we would need for the coming months (including my fiddle!) and we were finally ready to leave.

I was hot, bothered, and dripping with sweat, but Vlad was positively merry – either from the excitement of what lay ahead, or from the elderflower champagne, or perhaps a combination of both.

The day was scorching and there wasn't so much as a hint of breeze as our unlikely fellowship set off along lanes sunk deep between high hedgerows, which cast no shade in the blazing midday sun. Black asphalt baked in the heat and in places it had even started to melt into pools of thick tar that stuck to the horses' hooves.

I was leading Dakota because I didn't know how he was going to take to this new itinerant way of life; and frankly, after his reaction to the packs that morning, the idea of riding him was not an appealing one. To my surprise he seemed quite content, following quietly along

Setting Off

behind Oisín. Oisín, too, seemed pleased to be off on an adventure again, and was being his usual exuberant self: bursting into trot at random intervals, bouncing all over the road, and stopping to sample the hedges whenever Vlad wasn't paying attention – which was often! Even Spirit, my wolfdog, seemed happy to be on the move. Her traumas of the previous journey – during which Taliesin had trodden on her twice – were now a distant memory. Vlad was leading her from atop Oisín because I had my hands full with Dakota and his unpredictable, skittish behaviour. She trotted happily along on the cool verges beneath the hedgerows, dodging the melted tarmac and pulling Vlad's arm half out of its socket whenever she stopped to sniff the undergrowth or scent-mark the trail.

The miles passed slowly in a cloud of horseflies under the burning sun as we made our way down to Horsebridge where, only eight months before, Taliesin and I had crossed over the River Tamar back into Cornwall. Now Vlad, Oisín, Dakota, Spirit, and I crossed it again – this time heading out into the wide world to who knew where. What adventures awaited us? I felt a sudden surge of happy excitement. This was life! We were living the dream!

It was nearly eight o'clock in the evening and the temperature was only just beginning to drop when we finally arrived at Brentor on the western edge of Dartmoor. I was tired and aching after walking the best part of fifteen miles and already I had blisters appearing on my toes. Our host for the night, Caro Woods, gave us a warm welcome. After showing us to the field and leaving us to set up camp, she invited us in for dinner.

Caro was an artist and she lived in a converted chapel that also served as her art studio. The orange glow of the setting sun illuminated beautiful paintings which hung on the walls between tall, arched windows. Half-finished pieces rested on easels, or stood propped against the walls of the open-plan, top floor of the building where Caro had set the table for dinner.

Caro was in her sixties and although white hair framed her face, she had a young spirit and a real thirst for life. Three years earlier in 2015, she had made a pilgrimage from Lindisfarne off the coast of

Northumberland all the way to St Michael's Mount in Cornwall with her Connemara pony, Tommy.

A pleasant evening was spent sharing tales from the road and discovering mutual friends we had made on our respective journeys. It was midnight by the time Vlad and I crawled into our sleeping bags, and soon after I heard the horses slump down outside the tent and begin to snore softly. We were all exhausted.

The next morning, feeling a little more confident and still rather sore-footed from my hike the day before, I put Dakota's packs behind the saddle and rode him. Caro escorted us on Tommy, and her friend, Claire, walked alongside us for the first few miles. Dakota was jumpy and tense, and I was nervous, but he never once put a foot wrong. Even so, I hopped off and led him when we reached a stretch of open moorland, just in case. Best to quit while you're ahead, I thought, wincing with each step at my painful blisters and stiff, aching muscles.

With only eight miles to cover that day, we stopped in Lydford and took refuge from the heat at the pub. It had a large field out the back where we turned the horses loose for a few hours to graze. By four o'clock the heat was still stifling and oppressive, but we loaded all our gear back onto the horses and set off again. We followed narrow lanes flanked by high hedges full of horseflies which feasted on us relentlessly. For every blood-thirsty horsefly we squashed, another three would appear to take its place and so the journey became an epic, but utterly futile, battle against the irritating creatures. I felt not an ounce of remorse for the trail of bodies we left in our wake; the white hair on Dakota's neck and chest became red with the blood of his enemies, and he quickly became desensitised to my slaps and swats. If nothing else, it was good training for him.

Sticky with sweat and the blood of crushed horseflies, we arrived at the home of Liz and Guy – our hosts for the night. Liz had contacted me the year before to offer Taliesin and myself a stop on our way home from Scotland, but we'd already had somewhere to stay. When she'd offered us accommodation yet again, I couldn't refuse.

Liz was a keen rider with an interest in equestrian travel, and her husband, Guy, was a talented engineer with a passion for Land Rovers. After a much-welcomed shower we all sat out in the warm

Setting Off

evening sunshine enjoying a delicious home-cooked dinner. While Vlad and Guy discussed engineering and van conversions, Liz and I talked about all things related to equestrian travel. Finally we tumbled into bed feeling utterly content, and basking in the warm hospitality of two strangers who had fast become good friends.

This was definitely what life on the road was all about.

6

Meldon Viaduct

The next day we found ourselves on the Granite Way, an old railway line turned multi-user trail that runs along the western edge of Dartmoor between Lydford and Okehampton. We were just about to cross the Meldon Viaduct and here the solid tarmac of the trail gave way to wooden planks which reverberated loudly under the horses' hooves.

Dakota was not convinced. Every muscle in his body was tense and ready for flight.

Gingerly he took a step forward – one hoof, then two hooves – onto the bridge. His head was lowered, nostrils flared, and I could see the whites of his eyes from where I sat in the saddle, perched snugly amongst the packs. He felt like a tightly coiled spring, ready to take off at any moment.

Emboldened by Dakota's so-far positive response to his new role as a Road Horse, I had decided to ride him that morning when we left Liz and Guy. Now, surveying the hundred and fifty foot drop on either side of the low wooden railings that bordered the viaduct, I wondered doubtfully whether I'd been just a little too hasty in nonchalantly climbing aboard; and I questioned whether perhaps I would have been wiser getting off to lead him across.

Meldon Viaduct

But it was too late; I knew if I tried to dismount now, Dakota would certainly panic and bolt. I had no choice but to sit tight and hope for the best.

A third hesitant hoof went onto the bridge, and then the last. Dakota froze, petrified. Oisín, Vlad, and Spirit were already a good quarter of the way across the viaduct – Oisín plodding contentedly along, unperturbed by his resounding hoof-falls, and all three utterly oblivious to Dakota's predicament.

Then it was all too much. Dakota came unstuck, danced on the spot for a few seconds as he tried to keep all four feet off the unfamiliar surface before doing a one-hundred-and-eighty-degree spin and setting off at a flat out gallop back the way we had come. Two cyclists dived into the hedge as we hurtled past.

'Here we go!' I thought to myself with dismay as we charged along the asphalt at speed and horrific visions of the many possible outcomes flashed through my mind. What had I been thinking bringing this flighty young horse on such a crazy journey? And why on earth had I insisted on riding him that morning? He wasn't up to it; he was too scared of everything and his behaviour was too unpredictable. It was madness. There was no way this could end well; someone was going to get hurt.

Luckily, little more than two hundred yards down the trail, Dakota remembered that he didn't know where he was. He had nowhere to run to, and the only security and familiarity he had out here in the big wide world was Oisín – who was still on the viaduct, now neighing frantically for his friend, accompanied by Spirit's howls of concern.

Dakota slowed down enough for me to turn him around, and then he stopped dead, his whole body trembling with fright. Feeling a little shaken myself, but very relieved to still be in one piece, I quickly dismounted. I led the sheepish-looking Dakota back past the two terrified cyclists – who were still pressed tightly against the hedge – stepped resolutely onto the viaduct and with only a little hesitation, Dakota followed meekly. We made it to the other side without further mishap.

I decided not to ride again that day, choosing instead to walk the remaining sixteen miles to our next stop on my aching limbs and blistered feet.

We were heading to the Bowyer's that night. They had put me up the year before and I was looking forward to seeing them again. Emma Bowyer was a sensible, down-to-earth sort of person who knew a lot about horses, and also understood my passion for horseback travel. I was hoping that, as an equine behaviourist who regularly worked with 'problem' horses, she might be able to give me a few helpful tips for dealing with my nervous lunatic of a horse before we caused a serious incident. I was also hoping that Andrew, Emma's husband who was a well-respected master farrier, might be able to help us with a yet more pressing problem: Dakota's shoes – or rather, his lack thereof.

Dakota was still barefoot because shoeing him had been absolutely out of the question when the farrier had come the day before we left. He'd been much too terrified to let a stranger get near him, let alone pick up his feet, trim them, and nail hot shoes to them. I didn't have any hoof-boots for him either, so we'd simply set off and nonchalantly left it up to Fate in the hope that something would turn up. By the end of the third day of walking on abrasive tarmac Dakota was becoming foot-sore. Something needed to be done and as luck would have it, Andrew was home and willing to help.

When we arrived at the yard, we received a warm welcome of cold beer, and gin and tonic served by the pint. After we'd finished unburdening the horses, the moment I'd been dreading arrived. I had hoped that three days on the road and a good forty-five miles would have had a calming effect on my semi-feral young horse, but as Andrew approached, Dakota leapt to the end of his rope in full panic mode.

Clearly not, then! There was still absolutely no way that Dakota was going to allow himself to be shod. What on earth were we going to do now? We definitely couldn't continue with Dakota barefoot – another day on the road and he'd be lame.

Fortune must have been smiling on us that day, because just as we were starting to despair, a vet arrived. The timing could not have been more perfect. The vet had actually come to look at one of Emma's horses but he was more than happy to sedate Dakota for us.

After a minor scuffle in the stable, the vet managed to inject Dakota and half an hour later Andrew expertly set to work on my now very sleepy, totally relaxed horse. By the time the sedative wore off Dakota was wearing his first full set of shoes, and had also received his first hosing off for good measure. I didn't know what we would do when those shoes wore out, but I resolved not to think about it. We'd just have to cross that bridge when we came to it!

7

The West Country

For the first hundred miles or so I struggled to come to terms with Dakota and all his quirky behaviours. Most of the time I led him, but as the days went on, I gradually began to ride him more and more. By the end of the first week he had finally started to relax a little and settle into the journey. In spite of that, people and drains were still viewed with deep suspicion, and on several occasions he nearly decapitated the many thick-witted cyclists who sped silently up behind us and tried to overtake without giving us any warning of their presence. The offending cyclist would then usually shoot me a dirty look, as though somehow it was my fault that they had spooked my horse, and not theirs. Thus I grew to harbour a deep resentment towards all cyclists, and was eternally grateful to the small minority who had the good sense to call out from a safe distance and let us know they were there.

We retraced my route from the previous two years' adventures, staying with old friends and making some new ones along the way, too. Dartmoor's rugged tors gave way to rolling Devonshire countryside; sunken narrow lanes snaked up huge hills between sun-yellowed pastures where sheep and cows dozed in the shade of broad oak trees; and steep roads wound down into deep valleys where sluggish, dried up streams trickled away towards distant rivers. The whole world baked under cloudless, azure skies. The heat and the horseflies remained unrelenting and unbearable. Sweat dripped, horseflies bit, and we travelled in a state of perpetual discomfort.

It was so hot that finding water for the animals quickly became our biggest daily challenge. All the gullies, ditches, and puddles that usually lined the lanes had long since dried up, and the few rivers we came across were inaccessible – blocked off by thick hedges and wire fences. I observed with fascination how Oisín seemed to know exactly where to look for water, eagerly searching the ditches at the roadsides, and picking up the pace whenever we approached a river. His instincts were finely tuned and infallible – but to no avail in that heat. We resorted to sneaking into fields to let the animals drink from troughs, hoping the farmer didn't catch us, or asking at houses for buckets of water. On one occasion a couple who had passed us on their way home from work met us on the road outside their house with buckets of water at the ready. The kindness of strangers never failed to both humble and amaze us.

We took our first day off near Dulverton in west Somerset where Lisa Walker had once again offered us a place to stay. The house was being refurbished and no-one would be home, she'd said, but we'd be welcome to pitch our tent in the front garden and the horses could have one of the paddocks.

As we rested, the heat finally broke. The heavens opened, and it poured with rain. Sheets of water bucketed down onto the surrounding landscape and the steep lanes around Dulverton became gushing torrents of water, which tumbled down the hillsides to swell the sluggish rivers in the valleys below.

Late in the afternoon the rain stopped, the clouds dispersed, and by the time we packed up the following morning the heat had returned with a vengeance.

We picked our way through the rolling Somerset countryside along narrow, shady lanes towards Milverton, and then out along busy roads towards Taunton. That night we stayed at the Conquest Centre, one of the two charities we were supporting with the ride.

The Conquest Centre used horses to help people, offering a number of different equine assisted therapies, from therapeutic riding and driving for the disabled, to psychotherapy and equine assisted learning. I had stopped there for a night the previous year, bedding down in front of a roaring fire in the on-site yurt. Although my stay had been brief, I'd been struck by the ethos of the place and how the Centre valued the health and wellbeing – both physical and mental – of their horses. They put the needs of their equines first and foremost in their work, and their approach to horse care, horsemanship, and therapy was holistic and forward thinking. They had willingly helped me on my journey the year before, so it was nice to be giving something back, and helping them with some fundraising.

It was much too hot for fires and sleeping in yurts this time round. Instead – once the horses were settled in a field with a nice pile of hay and plenty of water – Vlad, Spirit, and I made camp in the Centre's forest school, in a small patch of woodland where their herd of therapy horses roamed freely. We cooked our dinner on an open fire and slept in hammocks slung between the trees, watching the summer stars through gaps in the motionless canopy overhead, and listening to the sound of horses moving quietly through the undergrowth around us.

Beyond the Conquest Centre the land levelled out as we made our way towards Burrow Bridge and the home of my dear friend April Anderson. This was the third year in a row that I had stayed with her – enjoying her warm hospitality and good friendship while the horses gorged on rich grass with April's lovely neighbour, Cheryl Green, who had once again kindly offered the horses a paddock for the night.

We took another day off there, and again it rained. I was glad not to be on the move then, because between scorching heat and pouring rain I definitely prefer the heat.

Vlad and I spent most of our day searching for his wallet, which he'd dropped somewhere on the road between Lyng and Burrow Bridge the evening before. It contained his driving license, all his credit cards, and worse still – his tobacco! Although we scoured the hedgerows and ditches, not a trace of it could be found. Eventually, we gave up and took refuge from the rain in a pub where Vlad began the lengthy process of cancelling all his cards.

The West Country

By early evening the rainclouds had dispersed and the sun shone hot once more. We climbed Burrow Mump – the hill behind April's little flat – and sat in the shadow of the ruined church watching the sun go down in the west. The Somerset Levels opened up below us, parched and yellow after many weeks of persistent heat, and in the distance we could just make out Glastonbury Tor rising out of the hazy, sun-scorched plain like an island in a sea of burnt gold.

8

The Somerset Levels

Dakota's ears pricked, his head came up, and I felt him tense at the sound of shrieking children and balls bouncing on asphalt. We were making our way through Othery, it was lunchtime, and there were children playing in the schoolyard next to the road.

There is something deeply alarming, even for me, about a group of six year olds who, on spying a horse in the distance, come flocking to the school fence shouting 'Horsey!' at the tops of their lungs.

I gripped the lead rope a little tighter as we approached the school, anticipating just how well this was going to go down with Dakota. As yet he had never experienced this ordeal so I was glad that I was walking and not riding at the time.

Wait for it, I thought, bracing myself as we neared the school. Wait for it...! And then:

'Horsey!' The shrill cry was taken up by a dozen eager children, just as we came level with the line of fencing that was more worthy of a high-security prison than a playground. Balls rolled abandoned across the schoolyard as hordes of screaming children left their games and came running over to the fence, pressing their eager little faces up against the wire mesh to get a closer look at the unusual sight.

The Somerset Levels

As anticipated, it was too much for Dakota. He shot sideways across the road, narrowly escaping a collision with some passing cars, and then shied wildly to avoid treading on a drain cover. It was as much as I could do to hang onto him as he flitted about on the end of the rope until we had passed the playground and the shrieks of delight subsided. It was a great relief to find ourselves back on quiet lanes again.

We set off across the flat Somerset Levels on straight, narrow roads flanked by tall willow trees and ditches of stagnant water – into whose reeking depths Spirit leapt with glee, showering us all with foul-smelling droplets.

Those of you who have read about my journey from Scotland to Cornwall in 2017 may remember that awful episode in Staffordshire when Taliesin trod on Spirit's paw and she had to be sent home. You will probably be wondering why on earth I'd decided to bring her along on yet another crazy adventure – and it's a fair question!

Spirit isn't a keen traveller. She definitely prefers a more leisurely existence of short walks, regular meals, and a comfortable bed to sleep in at the end of the day. I have to say I find her sense of adventure woefully lacking, and frequently accuse her of being an embarrassment to her species. What self-respecting wolf craves the comfort and security of home when there's a whole world out there to explore? Poor old Spirit, she's a long-suffering but faithful companion. After her ordeal the year before, I would much rather have left her at home – especially in that heat wave – but unfortunately it wasn't that simple.

Spirit is a neurotic creature who suffers with separation anxiety and has a horribly destructive streak. She's an escape artist who can open any door or window. Failing that, she can chew or claw her way through pretty much anything if she sets her mind to it; and to top it all off, she's a total liability with small animals, livestock, and other canines. She's not exactly the easiest creature to look after, and certainly not one that I would willingly inflict on anyone for any lengthy period of time. My close friend Cate is about the only person in the world whom Spirit seemed happy to be left with, but I couldn't have burdened her with the responsibility of looking after this neurotic lunatic for all the months that we'd be on the road. That would have been too much of an imposition.

In the end, we'd had no choice but to bring Spirit along with us and hope she didn't get trodden on again. And so far, so good! She stayed on a lead around livestock and small animals; a close eye was kept on her whenever we stayed with people who had dogs; and – barring one unfortunate incident with a spaniel – all creatures we met en route came out unscathed. Really, the worst part about travelling with a dog this time round was the insufferable heat. Finding shade, water, and dodging the many pools of melted tarmac was her biggest daily struggle. For the most part I let her walk where she chose – usually on the grass verges, or in the shadow of the hedges – and any time we came to a stream or a ditch where there happened to be water she would jump in for a nice, cooling swim. Whenever we stopped to let the horses graze, she would lie down panting in the nearest patch of shade to rest her weary paws.

A quiet day was spent crossing the Levels and it was late in the afternoon when we finally reached Glastonbury, where we would be staying with my old friend Guy Clifford. He had been a feature in all of my previous adventures, driving me to Scotland and coming to the rescue when Spirit was injured. I've always said that if ever you were stuck and in need of assistance, Guy would be your first port of call. A more genuine, kind-hearted person you couldn't wish to find and it was a pleasure to see him again.

Showers were offered and gratefully received, clothes were washed, and a nice evening was spent catching up on all his news. Exhausted, we finally tumbled into bed and slept soundly.

With only eight miles to go the following day, we spent a lazy morning exploring Glastonbury. We climbed the tor, drank from the springs, and looked around the quirky little town full of peculiar shops and strange and wonderful people. In the late afternoon we saddled up and set off across the levels again, heading for the Mendip Hills. There, nestled at the foot of the tall escarpment, lay Redmond Bottom Livery Yard: our stop for the night.

The Somerset Levels

From Redmond Bottom we navigated a maze of busy roads through the centre of Wells. Here the numerous drain covers and frequent encounters with people rendered the first few miles of the day a rather terrifying experience for poor Dakota, and I congratulated myself on having the foresight to be dismounted and leading him. He darted left and right, nimbly dodging drains and manhole covers, and snorting his displeasure as he warily eyed up the many pedestrians on either side.

All was going as well as could be expected until we suddenly found our way blocked by a metal grate, which ran from one side of the street to the other covering a shallow gutter in the road. Oisín ambled nonchalantly across it, barely even aware of its existence, but Dakota froze, terrified. A small audience of passers-by gathered on the pavement and a line of cars assembled behind him as Dakota danced before the metal grid, which was little more than six inches across. Finally, he geared himself up and sprang over the drain, all four hooves pulled tightly up beneath him, clearing the flat obstacle by at least three feet. Upon landing, he scooted quickly round to check that the drain was still in its original position, before hurrying to catch up with his friend.

Having somehow survived the ordeal, we soon found ourselves back on country lanes which led through acres of rolling golden arable fields. Here the roads were busy and cars rounded blind bends at terrific speeds, passing us just a little too close for comfort on the narrow roads. Much of the day was spent cursing inconsiderate drivers and thanking my lucky stars that, for all his faults, at least Dakota was sensible when it came to traffic.

At the end of a rather hot and harrowing day, it was a relief to arrive in one piece at our stop near Faulkland where the lovely Maggie Brombley put us all up for the night.

9

Settling In

After the first nerve-wracking few days, I finally began to trust Dakota in spite of his on-going irrational behaviour. I had come to realise that although nervous and skittish, he hadn't a nasty bone in his body. His spooks – which were frequent and dramatic – were always genuine. He never once tried to throw me off, and he always took me with him when he bolted – a fact for which I was eternally grateful. I was also grateful that amongst all this kerfuffle, with Dakota shooting off down the road at the slightest twitch of a leaf, or leaping six feet sideways whenever he was ambushed by a drain cover or an unexpected human, Oisín barely batted an eye-lid.

I fear I may have painted a rather unflattering picture of Oisín in my last book when I described how, on our journey to Land's End in 2016, he became so anxious that I couldn't be let out of his sight for more than a few seconds without him turning into a destructive maniac who would try to climb, paw, or barge his way out of any enclosure. I'd even had to sleep outside in the cold April rain next to his fence where he could see me at all times and travel home in the trailer to prevent him having a complete meltdown. While all of that was true, when in company Oisín was an altogether different horse. He was sane, sensible, and totally unresponsive to Dakota's frequent panics. He took a confident lead on our journey, assessed each situation for himself, and reacted as he saw fit – which usually amounted to stopping to graze the nearest patch of good grass in a vain attempt to satisfy his insatiable appetite.

Oisín is a large horse. Not tall, just wide. I think he's a Breton horse – a French draught breed descended from the warhorses of the Middle Ages. Oisín is definitely one of life's survivors whose sole aim in this world is to find – and eat – as much food as he possibly can. There is almost nothing that he won't do for alimentary reward, and on the rare occasion that something does frighten him, his anxiety can usually be appeased with a few minutes spent nibbling on some lush grass. Although horses like Oisín are hard to manage in normal, sedentary circumstances due to the ease with which they become over weight, they actually make excellent travel horses because they will eat almost anything without complaint, don't require much to maintain their condition, and they thrive on plenty of hard work.

I was so preoccupied wondering what on earth Dakota was going to do next, and seriously questioning my wisdom in taking this nervous lunatic of a horse on the journey, that I barely had time to think about anything else. Landscapes went largely unnoticed, and people were given only the bare minimum of attention. You try travelling with a nervous young horse and you'll soon know what I mean. It was all-consuming and utterly exhausting.

Vlad was proving to be an excellent travel companion and seemed to be adapting well to the rhythms of life on the road, but I feel it is only fair to let him share his own experiences of setting off on such a long journey:

> 'Wake up. Pack down. Saddle up. Load the horses. Ride until your arse hurts. Dismount. Adjust the slipping packs. Walk until your feet hurt. Try to ignore the blisters. Sweat buckets. Get back on. Sweat some more. Become dehydrated. Find water. Readjust the packs. Dismount and walk again. Find a stop for the night. Unload the gear. Turn the horses out. Set up camp. Take your shoes and socks off. Breathe. Cook. Eat. Laugh. Sleep. Repeat.
>
> 'It was physically and mentally exhausting!

'As romantic as it may sound, life on the road can be difficult. Although our needs were stripped down to the bare necessities – so much so that finding food, water, and shelter felt like significant achievements in themselves – I soon became overwhelmed by the number of daily issues that we encountered. Between the heat, the slipping saddlebags, finding water, and navigating busy roads, life was a far cry from the romantic dream of riding off into sunsets. Not to mention Spirit's constant pulling! What had once been funny and endearing about Oisin's rather jovial temperament and his stomach-driven mentality soon became something of a nuisance. If mounted I could barely control my horse; when leading from the ground he would pull me wherever he wanted to go – usually to the nearest patch of lush grass. If I was talking to Cathleen, or taking something out of my pocket, Oisín would take full advantage and stop dead in his tracks to put his head down and graze. He always seemed to know when I wasn't giving him my full attention.

'"Don't let him stop! Shorten your reins!" was Cathleen's constant cry of desperation.

'Even at three miles an hour, the day would pass quickly and we would once again be looking for our stop. When we'd sorted the horses out for the evening we were able to enjoy the company of our hosts, have a cold drink or warm meal, and put our feet up. Laughter, stories from the road, and fiddle playing in the moonlit camp were pure joy!

'The first hundred miles mark came fast and with it the realisation that anything is possible, with the right attitude and a sufficient timeframe. If I could ride a hundred miles, then there was no reason why I couldn't ride a thousand miles if I put my mind to it. I could finish the whole journey! It was tempting. There was no doubt about it: I was falling in love.'

10

Lumps and Bumps

'Oh my god! Look at these!' I said to Vlad with dismay, pointing to Dakota's belly. Two enormous swellings had erupted in the place where the girth had been before I'd removed the saddle.

We had just arrived at our stop for the night near Codrington following a particularly hot day crossing Bath, where finding water had been especially hard. The hills had been huge and the valleys deep, but the landscape was dry. Gone were the fields of livestock with their useful water troughs; instead we'd found ourselves wandering first through the tarmacked desert of the city, then through woodland, and finally over acres of golden arable fields with not a trough in sight. Since leaving the city, we'd found no houses where we could ask for buckets of water. For the first time on the journey, I'd become worried about the animals who were beginning to struggle with the heat.

It had been a relief to reach our destination. The horses were untacked, water was offered round, and we set about making camp for the night in the garden behind the house.

The relief was short-lived, however, when I discovered two hot, fluid-filled swellings underneath Dakota's girth. Both were about an inch in diameter. Perhaps an insect had bitten him, or maybe a seed had worked its way in and irritated the skin? Whatever it was, the lumps were there, and they were bothering him.

My heart sank.

I knew all too well that a rub could quickly become a swelling, a swelling could become a sore, and a sore could bring the entire journey to an abrupt and premature end – or at least put it on hold for several weeks!

Rubs, swellings, and sores are the bane of every equestrian traveller's life, and seem to be a near inevitable occurrence – especially in the blistering heat. Oisín had developed a swelling under the saddle behind his shoulder on the first day which had finally opened up into a small sore, and while it didn't seem to bother him, with all the heat and sweat it was taking forever to heal. Pads were cut and shims inserted in a futile attempt to stop the saddle from rubbing, but to no avail; so we treated it as best we could and kept an eye on it to make sure that it didn't become any worse. That was about all we could do, but it was disheartening all the same.

So far Dakota's skin was still intact, but if I saddled him up the following morning the girth would almost certainly rub him raw and exacerbate the problem, which we certainly didn't want.

Luckily our hosts, Hazel and Pete, were nice, understanding people and when we told them our dilemma, they invited us to stay for another night and see how things looked after a day's rest. The offer was gratefully accepted. Another day off wouldn't do any of us any harm.

By the next morning the swellings had gone down significantly and were no longer hot or tender to the touch, so we decided to hit the road again and see how things went. Erring on the side of caution, we added in an extra stop to shorten what would otherwise have been a very long day, and I placed a piece of soft sheepskin under the girth to prevent any further rubbing.

We made our way into the pretty market town of Chipping Sodbury and wandered down the high street in the heavy morning traffic. Dakota now dodged the drains and people with far less enthusiasm than before, giving them a wary twitch of his ear and a bulging eye, but no longer flapping about at the end of his rope in wild panic. His heart was definitely no longer in it, and I wondered whether it was finally dawning on him that neither were a threat to his immediate safety.

Beyond the town we found ourselves crossing a series of large, flat commons. Some were grazed by herds of cattle, and others were being cut and baled for hay. There were conveniently placed troughs of cool water on these commons from which the animals drank happily. Vlad and I, on the other hand, were stuck with our plastic bottles of water, which by ten o'clock every morning had become hot in the saddlebags and were horrible to drink. We were lucky if we could find anywhere to refill them during the day – but our concern was not so much for ourselves and our own comfort, as it was for the happiness and well-being of our four-legged friends.

The roads became increasingly busy as we neared Kingswood where we stopped at a shop for some snacks, and the horses shared a can of cold beer with us – eagerly slurping the cool liquid with their long pink tongues – before we pushed on to our stop near Wotton-under-Edge.

When we untacked him, the swellings under Dakota's girth were no better than they'd been that morning, but they were no worse either. They certainly didn't seem to bother him anymore, which I took to be a good sign, so we decided to push on and play it by ear. Sometimes that's the only thing to do.

Leaving Wootton-under-Edge, we followed busy roads across a high ridge in the Gloucestershire landscape for several miles until at last we turned off on a small lane leading into Horsley. Here the road dropped down into a deep valley, and we found ourselves on narrow, winding lanes which dipped and climbed through deep valleys and over tall hills as we approached Nailsworth and Stroud. Our destination was the Red Horse Foundation – the second of our chosen charities.

I'd stayed there the year before on my journey home from Scotland and had been very impressed by the work that they did using their small herd of horses to help people with a range of psychological disorders, mental health issues, and for personal growth and learning.

It fascinated me how horses could mirror humans and physically manifest underlying psychological or emotional problems, and I was amazed by how they could successfully be used to help clients work through and overcome those issues. Time and again it had been proven that profound results could be achieved much more quickly with the help of horses than through other, more conventional therapies.

When trying to decide which charities to support with the ride, it had been the first one that came to mind. The work they did was incredible, and you'd be hard-pushed to find a more deserving cause.

11

Tracks and Trails

There are two things you should probably know about me. The first is that I'm not a fan of the British rights of way system where walkers have one lot of designated paths, cyclists another, and horse riders are confined to a begrudging handful of tracks, which rarely lead anywhere useful and are often in a dreadful state of disrepair – if they exist at all. If there's an impassable bridleway to be found, then you can bet your life I've marked it on my intended route because I've got a real knack for that. I much prefer the Scottish system where anyone can go pretty much anywhere on foot, bike, or by horse – just so long as they're respectful of crops and livestock. It's a much more sensible way of doing things, if you ask me.

The second thing you should know about me is that, if nothing else, I'm pretty determined. Actually stubborn might be the better word to use here. I think you need to be if you want to travel long distances on horseback.

On the day that we needed to get through Gloucester and cross over the River Severn, my dislike for the English rights of way system got together with my rather pig-headed attitude and I settled on a more 'Scottish' approach to the route planning. That is to say, I looked at the map, and – upon finding no designated bridleways to get through Gloucester City – I decided to make use of the conveniently placed canal paths and cycle networks to get us through the maze of busy roads and over the River Severn. This was an appealing plan because it would save us a whole day's ride north to the next bridge, and another day's ride coming back on ourselves. I didn't see the point in adding all those extra miles and days onto our journey if we could cross the river in Gloucester. I also reasoned that anywhere a bike could go, a horse should be able to go too. I put this plan to Vlad, and he agreed it was a good one.

The day started well. We set off early from the Red Horse Foundation, picking our way along quiet lanes which gradually became busier and louder as we descended into Stroud and all the mayhem of rush hour on a weekday morning. There were buses and lorries, pedestrians and prams, cyclists and joggers, mums on the school run, and manhole covers on every corner. This was no place for a nervous young horse with an irrational fear of drains and people! But after two weeks on the road and more than two hundred miles behind him, Dakota had finally accepted that people were OK and that drains weren't about to leap out of the ground and swallow him whole. In fact he was so distracted by everything else that was going on around him that he forgot himself and even stepped on a few drain covers in passing.

The horses stayed calm and focussed as we picked our way through the noisy chaos, holding up the traffic and stopping frequently at lights and pedestrian crossings. All the same, I breathed a sigh of relief when we finally got away from the hustle and bustle of the busy streets, climbed out of the deep valley, and made our way along quiet lanes towards Gloucester.

As we approached the city, we had to cross the M5 for the second time on the journey. I braced myself, anticipating Dakota's meltdown. The first time we'd had to cross a motorway Dakota had panicked and tried to bolt at the sight of six lanes of heavy traffic disappearing under his feet – and who could blame him? Even normal, sane horses aren't too keen on that one. This time however, to my complete amazement, Dakota took the whole thing calmly in his stride – merely twitching a cautious ear and cocking a wary eye at the lines of cars and trucks moving beneath him. What was happening to my horse?

We wandered through large trading estates and housing complexes on the outskirts of Gloucester before picking up the canal path that led into the city centre. I had been hoping for easy access to the towpath and I breathed a sigh of relief when we found our way unobstructed. Even better, the path was wide, well surfaced, and almost deserted.

Vlad, who had been leading Oisín for a while, decided to mount from a mooring post next to the canal. As he did so, Oisín seized the opportunity to scratch his face on Vlad's legs – or perhaps it was an intentional head-butt? – and sent him flying to the water's edge. Another inch and Vlad would have found himself having an unplanned wash in the canal.

I howled with laughter at the sight of Vlad sprawled on the ground, muttering curses at his unruly mount. Impervious to Vlad's insults, Oisín immediately set to work eating the lush grass that grew along the banks of the sluggish, green waterway.

The canal path was easy going and no one stopped us, although we did get several quizzical looks because, strictly speaking, we probably shouldn't have been there. But since canal paths were actually created for horses, my sense of entitlement to their use appeased any qualms I may have had.

The canal ended in Gloucester docks and from there we picked up a nice, paved cycle path which ran through a large nature reserve. It was here that we finally came unstuck and the flaws in my somewhat haphazard planning began to show.

What I hadn't counted on, in my infinite wisdom, was the fact that the nature reserve was divided up into several large fields which were grazed by herds of cattle. The cattle weren't a problem, but the lines of stock fencing enclosing them were. The path we were following led across these fields and access was gained via kissing gates for pedestrians, and narrow cattle grids for cyclists. To the side of these were large metal five-bar gates – which were chained and padlocked.

My heart sank.

So much for horses being able to go anywhere a bike could! Why had I thought using a cycle track of all things was a good idea? I really should have known better! And why hadn't Vlad talked me out of it? What was the point in having a travel companion if they didn't talk some sense into you when you were having stupid ideas?

We had no desire to backtrack into town and pick up the A40, a busy dual carriageway, which was our only alternative means of crossing the Severn here. Nor did we want to detour to the next bridge several miles further north. We'd have to delay our pre-arranged stop, and hastily find somewhere else to stay for the night. It was too much of a palaver. No, somehow we would have to find a way through this nature reserve!

On closer inspection, we found that the five-bar gate could be lifted easily off its hinges to let the horses pass, but the next one we came to had inverted hinges so lifting it was out of the question. Feeling rather stumped, we looked around us. The solid wire fence ran unbroken across the nature reserve with no chance of undoing it anywhere to let the horses through; the kissing gate was an absolute impossibility; the cattle grid, however, was wide enough, and the metal bars close enough together, that in theory the horses could walk across it with ease – but how would we convince Dakota? He had barely come to terms with drain covers and manholes. Surely cattle grids were pushing it a bit!

Not one to be deterred by such trivialities, Vlad thought it worth a try. He rolled himself a cigarette for good measure, then – taking the lead – he pointed Oisín at the cattle grid and asked him to go forward. Oisín, ever the trooper, and no stranger to traipsing across cattle grids, clanged merrily over the metal bars. Next Spirit picked her sure-footed way over to the other side, and warily I followed. To my utter amazement Dakota crossed the grid with barely a hesitation. Could it be that he had finally turned a corner and was becoming the sane and sensible horse I had always hoped for? I hardly dared to think it!

We continued on over fields of dry, yellow grass, crossing more cattle grids until we reached the banks of the river. Here we found our way well and truly blocked by an underpass beneath a railway. The path was too narrow, and the bridge too low for the horses to get through.

My heart sank again.

Vlad refused to believe it and looked desperately for a solution. If we unloaded the horses and carried our equipment to the far side, could they squeeze through? Maybe they could crouch to get under the bridge? He suggested wildly, but I put my foot down. We'd had enough stupid ideas for one day, and there were limits even to my stubborn determination.

We backtracked a little ways over the fields, and while the horses grazed under my watchful eye in a patch of scrub, Vlad sallied forth to scout out an onward route.

'We'll have to scramble down a very steep bank, walk across a dried up bog, go under the road and the railway, and we'll come out by the bridge. There are a few gates, but they can be lifted off their hinges. Oh, and there's a narrow wooden bridge, but I think it's doable,' he said brightly when he returned from his recce some half-hour later.

I was dubious. Could Vlad really be trusted to know what was doable for the horses? After his earlier ideas about the underpass and his failure to identify my own stupid ideas I had my doubts, but we'd come too far to turn back now.

We set off following Vlad's directions through the seemingly impassable nature reserve. The slope was steep and the rock-solid clay of the dried out bog was pitted with many deep, ankle-breaking holes and cracks – all hidden beneath waist-high reeds – but the horses picked their way through carefully and all were unscathed.

Reaching the far side of the bog we passed under the dual carriageway, then under the railway, and finally we found ourselves confronted with a rickety wooden bridge, spanning a small ditch. At two foot wide, it was much too narrow for the horses to walk over.

'I think it's doable!' Vlad repeated, optimistically.

I did not; and nor did the horses when Vlad tried to persuade them onto the bridge.

Luckily, to the side of the footbridge was a wide gate – but it was padlocked and its hinges were all tied up with tightly twisted wire. Vlad set about undoing the wire with a pair of pliers and a short while later the gate was lifted, our horses led through, the gate replaced, and at last we found ourselves on the bridge crossing over the Severn. We took a moment to breathe and admire the views before pushing on again to see what further challenges awaited us.

The cycle track eventually brought us out on a pavement, which ran alongside the A40 dual carriageway facing into the oncoming traffic, before finally veering off across open farmland to Highnam.

I was glad when we found the solid tarmac of quiet roads under our feet once more.

I decided in that moment that perhaps using cycle tracks wasn't such a good idea after all, and I also concluded that Vlad could definitely not be relied upon for sensible ideas or the provision of a much-needed voice of reason.

12

The Forest of Dean

One important thing that I've learnt on my adventures over the years is that local knowledge is not all it's cracked up to be, and that any advice people offer with regard to routes should be received with a certain amount of scepticism. I find it is best to go with my own gut instinct about routes – except in the case of cycle paths! – and to take any proffered suggestions with a healthy pinch of salt. I have arrived at this conclusion not because I think I know best, but because people frequently provide highly inaccurate information about the state of their local roads and bridleways. What some people consider far too dangerous to ride can be a walk in the park, and what others consider a good idea can be a total nightmare – as in the case of one equestrian who tried to direct me down a busy dual carriageway bypass near Frome, convinced that it was a perfectly safe and sensible option!

Whenever anyone offers me advice, I have to quickly work out whether or not the information they're giving me is accurate and reliable, and whether I should take it or not. This evaluation process is done subconsciously using roughly the following criteria:

Who is offering the advice and what sort of person are they?
Do they seem to know what they're talking about?

Are they an equestrian or not?

If they are, then what is their preferred discipline? (i.e. jumping or dressage, happy hacking or endurance?)

Are they a confident or a nervous person?

What kind of horses do they ride?

The gist of what I'm getting at is this: as a rule, I won't listen to anyone's evaluation of the state of a bridleway or the local lanes if they have no idea about horses, or if they try to send me down busy dual carriageways at rush hour. If they are an equestrian but all they do is ride in a sand school on a flighty Thoroughbred who's scared of his own shadow, I rarely put much store by what they tell me either, because to them most roads are deemed too dangerous to ride and the majority of bridleways too full of potential hazards. I'm much more likely, however, to heed the advice of someone who hacks out on a calm, sensible horse, and who appears to really know the local area and all the routes in it. Unfortunately even this finely honed system is not infallible, as we discovered when we got to the Forest of Dean.

Vlad and I had arrived at our stop feeling tired and fed up after the Gloucester cycle track fiasco and eleven hours spent wandering about in the baking heat. A much-welcomed shower and a hot meal with our friendly hosts were enjoyed, before we bedded down on the concrete floor of their enormous barn for a good night's sleep. In the morning after breakfast, we settled down with mugs of steaming coffee to study maps of the Forest which were spread across the kitchen table.

'The best way to get into the Forest is here,' said our host, Geoff, pointing to a track marked on his OS Explorer map. It wasn't a designated bridleway – just a footpath leading into Flaxley Woods on the north-eastern edge of the Forest of Dean.

After the debacle the day before, my faith in using anything that wasn't clearly marked as a road or bridleway could only be considered shaky at best, and I said as much.

'It's fine! We've done it hundreds of times!' Geoff assured us, confidently. His wife Carolyn agreed.

'There's a gate there,' he continued. 'You can't miss it. Just follow the track on the other side. It's all open and easy going after that.'

The Forest of Dean

Our hosts had lived in the area for over twenty years. Geoff had been the master of the local hunt, and they frequently organised fun rides in the Forest with their friends where they hacked for many miles. Not only that, but they had ridden in several countries around the world, too. They were horse-people to the core!

Knowing my system when it came to taking advice from strangers, Vlad gave me a meaningful look, to say he felt that our hosts met all the criteria for sane, sensible people who knew what they were talking about.

These were definitely the kind of people whose advice was not to be sniffed at, I decided, putting aside my doubts. It would be fine! So we marked down the route they suggested and set off.

The first few miles were uneventful as we rode along narrow, shady lanes which provided some respite from the insufferable heat, and after a few hours we reached the gate which our hosts had said would allow us access to the Forest. To our dismay, we found it locked.

We wandered around looking for an alternative way into the woods but found nothing. Eventually Vlad persuaded me that the best course of action was to simply cut across country in a straight line and hope for the best. Against my better judgement I agreed, and we dived into the nearest open patch of woodland.

After less than a hundred yards we came up against a line of live electric fencing. Rather than turn back and take a long detour to the next obvious access point into the forest, we managed – with several electric shocks and lots of cursing – to drop the fence and get the horses safely across it. Once the fence was restored, we found ourselves on an old dismantled railway. Now our onward journey was barred by a line of thick wire fencing strung tightly between solid concrete posts and I was beginning to regret going along with Vlad's rather reckless plan. He had a total disregard for fences or private property in the countryside – two things which were almost unheard of in his native Romania, he informed me. But this wasn't Romania, this was Britain; fences abound, trespass is frowned upon, private property is fiercely protected, and pretending those things didn't exist hadn't exactly worked.

Fiddler on the Hoof

The track we wanted was barely two hundred yards away from where we stood, yet getting onto it was proving a total nightmare. Even so, backtracking was out of the question. We were committed now.

We wandered up and down the old railway line searching for a way through into the Forest, and eventually I spied a gap in the fence at the bottom of a very steep slope, beyond yet another line of electric fencing.

While Vlad watched the animals, I carefully removed the electric fence, scrambled down the steep bank, undid the wire that was tied loosely across the gap at the bottom, and then returned to fetch the horses. Vlad and Oisín went first, going slowly down the steep bank to avoid ripping open the saddlebags on low hanging branches. Oisín scooted and slid the last four feet to the bottom and squeezed carefully through the tiny gap in the fence without ado; then Dakota followed.

I can honestly say I have never been more proud of those horses. It never ceased to amaze me how willingly they followed us, and how uncomplaining they were of all the strange and downright stupid things that we asked of them. I was also grateful that Dakota had finally turned a corner and had stopped being a total lunatic, because we'd never have managed any of it otherwise!

We returned all the fences back to their original state and, after another scramble through the dense undergrowth, we finally made it onto the elusive track. As promised, the going from there was open and easy, but that day I vowed never to listen to anyone's advice against my better judgement ever again.

It was refreshingly cool under the trees and the Forest around us was silent, save for the singing of birds. We rode along wide tracks under tall stands of dark conifers, then through acres of native woodland where sunlight filtered through the thick canopy, illuminating the forest floor and casting dappled shadows across the trail. Sometimes our path led through patches of open heathland where flowering heather and sunburnt bracken grew thickly, interspersed here and there with lone oak, birch, and hawthorn trees.

The Forest of Dean

Dakota was happy and relaxed away from the traffic. He picked up the pace and took a confident lead – his long, quick strides soon leaving Oisín, Vlad, and Spirit far behind us. Spirit objected to this and whined loudly because she has an unhealthy obsession with being near me at all times, and Oisín objected, too, because he fancied himself the leader of our little troupe. He let out an angry snort and came thundering up behind – his heavy frame making the ground shake – then he barged past us, positioning himself squarely in front of Dakota in such a way that Dakota couldn't pass him. If Dakota so much as tried, Oisín would push him back into line with his enormous, chestnut backside.

The Forest stretched on for miles around us, a maze of criss-crossing tracks – some stony, others soft and covered in a thick layer of turf or well-rotted leaves. It was perfect for a canter! I'd never cantered Dakota – not intentionally, anyway. How nice it would be to gallop along the shady paths of the Forest unencumbered by our heavy packs, I thought wistfully.

Our stop that night was at Perrygrove Railway, an award-winning tourist attraction near Coleford where Katherine and David Nelson-Brown had kindly offered us a field. They were laid-back, cheerful people and they said it was no problem if we stopped there for two nights and spent a day exploring the Forest.

We woke up late the following morning and made our way up to the railway just as the first visitors were starting to arrive. Vlad couldn't believe his luck at finding himself in a steam-engine theme park, and could barely contain his excitement.

We queued up with the crowds of young children eagerly waiting to board the train, and chatted with some of the friendly staff who were as passionate and knowledgeable about steam engines as Vlad was. It wasn't long before Vlad had secured himself a seat next to the driver, and to his utter delight he even got to drive the train along the winding track, stopping to let visitors on and off at some of the little stations along the way and blowing the loud whistle with glee. He was like a kid at Christmas and he wore a grin for the rest of the day!

In the early afternoon, having finally managed to drag Vlad away from the trains, we saddled up the horses.

Fiddler on the Hoof

By this time the lumps under Dakota's girth were disappearing, and even Oisín's pesky sore had finally started to heal. We were happy, and all was right with the world as we headed back into the cool shade of the Forest, following tracks at random and stopping to feast on wild raspberries that grew in abundance along the edge of the path. The horses were keen and alert, and with no heavy packs to carry it wasn't long before we'd upped the gears.

Dakota was hesitant at first, unsure whether cantering with me on his back was ok, but he soon relaxed and gave it his whole-hearted best.

He had a beautiful canter, fast and smooth – it felt as though he were floating over the ground. My heart soared as we raced together through the silent forest, his little hooves beating the soft green turf while Oisín charged along in our wake, trying desperately to keep up with his graceful friend.

I'd never dared dream this day would come. My God, this horse had come so far!

13

New Shoes for Dakota

'It's a beautiful view,' I remarked.

'The best view in the world,' Morgan said with absolute certainty.

We were sitting on a bench in Morgan's field swigging from bottles of ale and watching the sun go down over the distant Black Mountains.

'I've travelled a lot but never found anywhere nicer. Wales is the best place on earth,' he said, contentedly.

We'd crossed the River Wye at Monmouth earlier that day and, after several miles of traipsing about on lanes through undulating countryside, we'd begun looking for somewhere to camp for the night.

Up until this point we'd arranged all our stops ahead of time, but now we decided to wing it and hope for the best. It gave us more freedom and flexibility, adding a nice element of spontaneity to the adventure.

Two young men, who had stopped to ask what we were doing, directed us to a bungalow a little further down the road when we enquired about somewhere to stop. There was a field behind the house, they said, and Morgan – the man who lived there – would probably let us camp.

Fiddler on the Hoof

We met Morgan outside the bungalow, and Vlad lost no time in informing him that we were looking for a field for the night, that we'd been told he had the best one around, and could we camp there?

It wasn't the approach I would have taken. Morgan – a tall, skinny man in his early thirties – looked a little taken aback at two random strangers turning up on his doorstep with horses and a wolf in tow, practically inviting themselves to spend the night with him. It was Vlad's first attempt at finding a stop and he'd not yet perfected the subtle art of getting people to offer us a field of their own volition, rather than forcing ourselves on them.

Still looking confused, Morgan showed us the acre of overgrown grass behind the house and said we were welcome to stay if we liked. After unloading the horses and turning them loose, Vlad set about making camp while I set about trying to find a farrier.

The moment I'd been dreading had finally arrived: Dakota needed new shoes. That morning when we left Perrygrove Railway and the Forest of Dean, I'd noticed that the toes of Dakota's front shoes were disintegrating rapidly. In the last week Dakota had really turned a corner: drain covers were no longer a source of terror for him, motorways were no problem, even people didn't seem to worry him quite so much. He was much more relaxed to ride and he didn't bolt off quite so often, either. A new set of shoes, though – that might be a step too far!

Finding a farrier who answers his phone is one thing, and getting one to come out at short notice is another thing still – but finding a farrier willing to come out to a horse that might potentially need sedation…?

'You can forget it!' I thought. But I had to try!

I rang a few numbers for local blacksmiths that I'd found using the online farrier registry. As usual, no-one picked up. I left messages explaining the situation, knowing full well that they'd never be answered. Then at last one John Crofts answered the phone.

'Hi, is that John?'

'Yes, speaking.' The voice was measured and slow, with a heavy West Country lilt, tinged with a bit of Welsh.

New Shoes for Dakota

'I'm riding my horses from Cornwall to Ireland. One of the horses needs new shoes. I was wondering whether you'd be able to put a new set on him?'

'Right,' he said slowly. 'Where are you?'

'Somewhere north of Raglan,' I said, scrambling for my map.

'What? Where did you say you were riding from?' It had just hit him.

'Cornwall,' I repeated. 'The horse might be difficult to shoe, he had to be sedated last time.'

'And where did you say you were going to?' John sounded genuinely confused and didn't seem to have heard the important bit about Dakota being difficult.

'Ireland. Do you have any sedatives just in case? Or maybe even a twitch?' I said hopefully, trying to bring him back to the real problem at hand.

'What, so you're camped at the side of the road now?' Again, he was missing the point here.

'No, we found a field. It may be impossible but could you come and see what you think? I'll still pay you for your time.'

We were having two conversations simultaneously, and John was definitely focusing on the wrong one!

'OK. Can I ring you back? I've just got to make a few phone calls. Cornwall to Ireland did you say?' He sounded incredulous and I wondered whether I'd put him off with all this talk of sedatives.

Apparently I hadn't, because within two hours John arrived. He was a slight man who looked to be somewhere in his sixties. He had short grey hair, intense blue eyes, and had a nice, quiet way about him.

I led Dakota to the gateway where John had parked his van. I was feeling apprehensive but trying not to let it show in case it made Dakota edgy. What would we do if he decided he was having none of it?

John approached slowly and gently stroked Dakota's neck, before bending down to pick up his hoof. To my utter astonishment, Dakota didn't seem in the least bit worried.

John handled Dakota in a soft yet confident way, as he deftly pulled the worn shoes off, rasped the hooves flat, then put a new set of shoes into the furnace to heat up. Dakota remained uncharacteristically calm throughout, barely flinching when John pressed the hot irons against his hooves, beat them noisily into shape on his anvil, and cooled them off with a hiss of steam in a bucket of water, before nailing them into place while Vlad handed him his tools.

I stroked Dakota's neck throughout, talking to him in a low voice and he soon fell into a dazed, half-hypnotised state with his bottom lip sagging and his eyelids drooping. The mood was contagious and I began to feel quite spaced out as well. There was something about the atmosphere that was very soporific and relaxing, and all the while John chatted away as he worked, stopping every now and then to stroke Dakota and tell him what a good boy he was.

I thanked Providence and the Universe for sending this lovely, gentle man our way because I sincerely doubted that anyone else could have done the job. He was exactly who we needed, where we needed him, at the time that we needed him most.

New shoes tightly in place and Dakota back in the field, I hugged John and thanked him from the bottom of my heart for all his kindness and patience. He shrugged, saying there was no need to be rough with horses when a quiet pat and some calm reassurance went a whole lot further. It was a simple philosophy, but an effective one.

I waved John off down the road and looked over to where Dakota was now grazing peacefully in the field with Oisín. I felt something in him had changed that evening. Gone was the wild, fearful look in his eye to which I had been so accustomed; in its place was a softer, more gentle, confident expression. What had happened to my horse? Where was the half-crazed lunatic who flapped wildly about at the end of lead ropes with his eyes popping out of his head at the mere suggestion of a stranger approaching? Who had swapped him for this calm, relaxed, and sensible creature?

Feeling relieved, elated, and still rather spaced out, I joined Vlad and Morgan on the bench looking out across the valley at the setting sun and admiring the view.

'That was amazing,' I said, turning to Vlad, still unable to believe how well Dakota had behaved. 'John was so good with him!'

'Well, he's a Welshman,' Morgan offered, matter-of-factly. 'What do you expect?'

I wondered what being Welsh had to do with it.

'The grass is a lot greener this side of the river,' Vlad observed after a brief, awkward silence. 'In England it was all brown and burnt, but as soon as we crossed the river it was noticeably greener.'

'Ah, Welsh soil, you see,' Morgan nodded sagely.

Conversation with Morgan wasn't really a two-way thing. It was more a series of statements on his part, to which we could respond as we wished.

'Scotland? Why would you want to go there?' Morgan said in disgust when conversation turned to my last adventure.

'It was beautiful,' I said. 'Have you ever been?'

'Nah.' He shook his head. 'It's full of Scots.'

I couldn't argue with that, but I failed to see how that was a bad thing. I'd found the Scottish to be a friendly and hospitable people on the whole.

'I like Wales,' I said, switching to a topic that I hoped he'd find more agreeable, wary of the way he was nervously flicking his folding pocket knife open and closed in his hand while he spoke.

'It's the best place in the world,' he said again, with conviction. The knife snapped shut.

'I went to Snowdonia a few years ago,' I offered. 'That was beautiful.'

'Nah. Don't bother with North Wales,' Morgan said, flicking his knife open again. 'They're all unfriendly up there ... and they speak Welsh,' he added as an afterthought.

Morgan was a builder by trade – a job he said he didn't mind – and he'd recently been converting the basement of the bungalow where he lived with his parents. They'd gone away for the weekend, he said. They'd be sorry to have missed us.

Vlad and I exchanged a look and an unspoken thought passed between us. For a brief, uncharitable moment, as we watched him flicking his knife – open and closed, open and closed – we wondered what was in his basement, and where his parents had gone.

14

The Black Mountains

Travelling in company was proving an interesting experience. On all my previous journeys I had travelled alone, save for my faithful animals; my days had been spent wrapped up in my thoughts, and drinking in the surrounding landscapes in blissful solitude. With Vlad at my side I not only had my own needs and moods, and the needs and moods of my animals to consider, but I now had to take into account those of another person, too. Gone were the long hours of silent introspection, the deeply meditative states induced by the slow rocking motion and rhythmic hoof-beats of my horse; instead, they were replaced by a ceaseless flow of noise as Vlad chattered happily on about this thing and that, always with a hyper enthusiasm.

While it was nice to share the experiences of the road – the daily challenges, responsibilities, and the many highs and lows that are unique to horseback travel – I have to confess that I sometimes caught myself longing for just five minutes of the peace and quiet I had experienced on my last journey.

After the first few days travelling together, we quickly fell into a routine and the daily chores were divided according to our individual strengths. Vlad was responsible for the more domestic tasks such as setting up camp and cooking, while I tended to the animals and dealt with the route planning and navigation. This worked for the most part because Vlad was an excellent cook and I knew the horses and understood their needs; I also possessed good map-reading skills, and I had a reliable sense of direction – which Vlad did not. Every now and then Vlad would complain that he never knew where we were going, or even where we had been, although the latter was due to him paying no attention to the places through which we passed and the former was because he never bothered to ask, so I refused to hold myself accountable.

Spending twenty-four hours a day, seven days a week, for months on end with another person is no mean feat, and it will test the strongest of relationships even under normal circumstances – never mind travelling hundreds of miles across country on horseback in the blistering heat and having to cope with the daily challenges and anxieties to boot! But as travel companions go, Vlad was a pretty good one. He was a laid-back, sociable sort of person, with an unshakeably positive and optimistic approach to life; little ever seemed to worry him. I am definitely not the easiest person to travel with – a fact to which I'm sure Vlad will happily attest! Therefore I was fortunate that he was so easy-going and good-natured, and most things just rolled off him like water off a duck's back. He was tolerant of my moods – which were largely dictated by the health and happiness of my horses, and the frequency and availability of my meals – and he did everything in his power to keep me in good humour by providing me with plenty of food, and a comfortable place to sleep at the end of each day. He was very creative when it came to rustling up delicious meals with a limited number of ingredients and we rarely ate the same thing twice, in spite of our unvarying staples of pasta, rice, and couscous – which we supplemented with fresh vegetables as often as we could get hold of them. As a result of Vlad's cooking I didn't lose so much as a pound on our journey, despite all the walking I was doing!

Vlad wasn't cheerful all the time, however, and on the few occasions where he succumbed to a bad mood – usually due to something I'd said or done – then he would sink into deep silences and barely say a word for hours on end. This would infuriate me because, although there was peace and quiet at long last, it was an uncomfortable, brooding kind of quiet that brought no solace, nor did it sooth the soul.

When he was sulking Vlad wouldn't cook. This was probably to punish me for whatever offence I had caused because, like Oisín, food is one of my main motivators in life, so in this way we would both fester for hours in our respective grumps until one of us – usually Vlad – broke the awful silence and amends were made. For the most part, however, peace and harmony reigned.

Our first argument happened about three weeks into the journey as we were leaving Abergavenny in Wales. It is amazing, under the circumstances, that it hadn't happened sooner, and it says a lot more about Vlad's patience and tolerance than it does about mine.

What sparked it was a minor disagreement about Vlad's insistence on riding in the middle of the road on narrow, winding lanes where cars came hurtling around blind corners at speed, along with his refusal to pull over and let vehicles pass.

Vlad maintained that by positioning himself in the middle of the road he could be seen sooner by approaching drivers, thus forcing them to slow down, but I genuinely feared both for his safety and for Oisín's.

Vlad can be pretty obstinate at times, and on this day he'd had enough of speeding vehicles, and enough of backtracking and pulling in for the traffic every two minutes. While I was fed up of it too, I was trying to be courteous and not end up in a heated debate with a pissed off driver – especially if we were only a few feet away from a convenient gateway. But on this day Vlad was having none of it.

'They can go back!' he insisted as we encountered yet another string of cars. We were barely three yards from a nice wide passing place.

'To where?' I asked incredulously. There was nowhere for them to go back to! 'We can go back!' I insisted, taking a more reasonable approach.

'No.' Vlad said flatly. Reasonable was not on the cards that day!

So he pushed Oisín on and squeezed past the stationary cars, while I turned and headed back to the wider section of the road. Because Dakota was heading in the opposite direction, Oisín grew anxious and attempted to turn round and follow; and Spirit – who Vlad was leading at the time – began to howl shrilly. When the cars had made it through the mayhem things descended into a blazing row, at the end of which I told Vlad to go home.

Vlad didn't go home, but silence descended like a black cloud and barely a word was spoken for the rest of the day.

We climbed steadily up a long road that led through Mynydd Du Forest. It was pleasantly cool under those dark pine trees, and silent – save for the steady clip-clip of the horses' hooves. The road eventually petered out to a rough, stony track that brought us onto the open hillsides and we made our way towards Grwyne Fawr Reservoir, whose many-arched dam and solid tower built in dark stone stood out dark against the backdrop of rolling hills.

Someone had said there was a bothy 'at the end of the reservoir'. I'd not thought to ask which end because they'd assured me that it couldn't be missed. Except that it could be missed, because we missed it – several times in fact – and it wasn't marked on my map, either.

The evening was drawing on, we were all tired and hungry, and finding somewhere to stop had now become paramount.

We wandered up and down the reservoir several times looking for the elusive bothy. Finally we headed across the dam and over the near-dry spillway, but there was no bothy to be found here either – only a lot of signs which read: 'Private – keep out!'

The grass was lush and Vlad was all for pitching our tent, in spite of the numerous private property signs, but the hills were overrun with herds of free-ranging horses and an encounter with an over-protective stallion could have been disastrous.

In the end, we decided that the best thing to do was to backtrack for a mile or so to the last house we had passed and see if they could help. There was a large public car park full of long grass there, which would make the perfect campsite if we could just find some buckets of water for the animals.

I knocked at the door of a pretty little cottage next to the car park. The lady who answered looked somewhat taken aback. After a brief conversation about who we were, where we were going, and what we were doing, she offered the horses the paddock behind the house and showed me and Vlad to the shepherd's hut in the garden. It was beautifully restored and kitted out with a double bed, electric sockets, and Internet. It was an unexpected luxury! Yet even having somewhere safe for the horses and a comfortable bed for the night did little to break the still-strained atmosphere between me and Vlad. He went straight to bed, and I cooked my own dinner that night.

The following morning tentative reconciliations were made, differences of opinion with regards to traffic management were put aside, and the heavy mood finally lifted. We readied the horses and made our way back past the reservoir and up over the mountains, before beginning a long, steep descent the other side. Here the landscape opened up below us – miles of gently rolling farmland, broken up by neat lines of hedgerows and a smattering of tall trees. Little farms and small clusters of houses were nestled amongst the patchwork of fields that stretched away to the north and west as far as the eye could see. To the south lay the Brecon Beacons, hazy in the baking heat, their parched slopes desperate for rain. Looking back at the Black Mountains as we picked our way towards Talgarth we saw tall columns of dark smoke rising from the hillsides behind us. The mountains were on fire!

While we humans were experiencing a daily series of emotional highs and lows, good moods and sulks, peace-making and arguments, the horses were much more stable in their humours, and they never seemed grumpy at all. They thoroughly enjoyed being on the move every day, exploring new places and sampling all the different grass that Britain had to offer – and beer, too, for which they had both developed quite a taste! It amazed me how they would come to the gate each morning, eager to hit the road and see what lay over the next horizon, and they seemed positively bored on their days off. For all the exercise they were getting, like me they didn't lose any weight at all.

If spending all day every day for weeks on end with another human is a challenging feat, then doing the same with a horse is entirely the opposite. There can surely be no nicer way to spend every minute of one's time than in the company of an equine – especially one whom you have raised and trained yourself. On the road, you get to see a very different side to your horse's nature. No longer is he a highly expensive pampered pet, milling about in a field and eating a hole in your pocket; instead, he is an equal partner on an adventure into the great unknown, carrying you or walking beside you, unfaltering, every step of the way. Over the many miles, strong bonds are forged with your four-footed companion, trust is built, and a new respect gained for his strength, stamina, courage, and absolute willingness to follow you against all the odds. It is both delightfully beautiful, and deeply humbling.

Now that Dakota had finally calmed down, I was discovering that under his distrustful, flighty exterior a sweet, affectionate little horse had been waiting to emerge. He still had his quirks, but they were no longer dangerous, over-exaggerated reactions to the world around him. Instead they had given way to a host of rather endearing little idiosyncrasies, like they way he would indicate with a flick of his tail or the swing of his head where a fly was bothering him that he wanted me to remove – IMMEDIATELY! Or the way he had worked out how to tell me – none too subtly – when he was tired and wanted me to get off and walk. After several miles of being ridden he would suddenly stop dead in his tracks and turn his head to look at me from the corner of his eye. If I pushed him on, he'd willingly oblige, only to stop again a few minutes later. He'd repeat the exercise until I dismounted and walked and it wasn't long before he had me well trained and responsive to his demands.

Oisín, on the other hand, was a stoic little horse and he'd carry his rider all day without complaint, never letting on when he was tired or wanted a break – except to eat! But we soon noticed, to our amusement, that whenever I got off Dakota to walk, Oisín would look round at his friend and snort loudly with a note of desperation in his voice. It happened every single time without fail and was almost as though Oisín were saying: 'Oi! My back's aching. This lump won't get off. How did you get rid of yours?' and Dakota would answer him with a sympathetic nicker.

Whenever this happened, Vlad would also dismount to give Oisín a break, and so we'd all walk for a few miles.

15

The Brecon Beacons

Beyond Talgarth, we picked up a series of bridleways that could barely be called such because they'd clearly not been used in a very long time. We cantered the first part along a deceptively pleasant, grassy farm track, which gave way to a narrow, overgrown path where thick vegetation came up to the horses' shoulders and the track underfoot became a boggy waterway. Crossing the busy A470, we lost the bridleway altogether and had to ask a man in the nearest house. He pointed us up his drive, past rusty, rundown cars and tall stacks of old fridges and washing machines. The bridleway went that way, he said, but it was all over-grown now and he'd not seen anyone on it for thirty years. We bypassed it, traipsing across sheep fields where all the gates were off their hinges and tied up with large quantities of baling twine, until we finally fetched up on a road.

It was getting late, so we decided to start looking for somewhere to stay.

Near Llanddew we stopped at a farm and got chatting to the farmer, Elwyn Price, and I asked him about somewhere to camp with the horses. At first he viewed us with mild suspicion, scratched his head, 'ummed' and 'ahhed' and said he couldn't think of anywhere, but after a pleasant chat – during which he had time to decide that we were genuine people and weren't about to rob him blind or murder him in his bed – he offered us an enormous field opposite the farmhouse.

We stopped there for two nights to rest the horses, and Vlad and I spent the day exploring Brecon – wandering the narrow streets, looking in the shops, and seeing what was on sale in the busy indoor market.

By this time my walking boots were starting to disintegrate; the sole was coming away from the boot, so we went in search of a shoe repair shop, too. The prognosis wasn't good, but since I couldn't afford new boots, we took a gamble and had them bodged back together again.

As we walked back to camp late in the afternoon, the heavens opened, rain bucketed down, and we got a thorough soaking – but the fires in the Black Mountains behind us were finally extinguished. So far the weather seemed to be working in our favour and saved all its heavy rain for the days when we weren't on the move.

By the next morning the sun shone hot, the land steamed, and everything dried out as we prepared to hit the road again. We bid farewell to the Price family – who had turned out to be a nice, friendly lot once their suspicions had worn off – and leaving Llanddew, we headed southwest towards the distant slopes of the Brecon Beacons. Although it was a two-day detour from our otherwise westward course, Vlad had insisted we ride across the hills and I'd had no objections. There's nothing I love more than travelling through the silent mountains, far away from busy roads and the chaos of civilisation. It soothes my soul, quietens my mind, and appeases my insatiable longing for wilderness and wide-open spaces.

A quiet day was spent meandering along little lanes through the rolling Welsh countryside until we reached the foot of the Beacon Beacons, where we found a field to camp in for the night beside a working farm.

In the morning, looking to refill our empty bottles, we asked a couple of farm workers for some water. They said the mountain spring – which the farm relied on for water – had dried up and there wasn't a drop on site, so we set off with empty bottles and hoped for the best.

We followed a wide, stony track for several miles which skirted the foot of the Fan Frynych, and then climbed sharply to an open sweep of barren moorland. Out here on the hills, far away from people and traffic, Dakota relaxed. He was in his element and totally at ease, striding happily along and leaving Oisín, Vlad, and Spirit far behind to pootle along at their more sedate pace. Thus the hours passed in happy silence as we rode through the beautiful landscape.

After a few miles, we found ourselves on a small road which we followed a little ways before picking up another bridleway towards Coelbren. By this time Vlad and I were parched with thirst, and every stream we came to made us ever more desperate for a drink. We'd still found nowhere to fill our water bottles and we were both becoming rapidly dehydrated. At last we came to a small bridge spanning a stream and a little ways down the hillside I spied a farm. Leaving Vlad to look after the animals, I ignored the many signs warning people to keep out and made my way down into the farmyard. I was greeted by a mangy dog which barked and growled at me ferociously as I knocked on the door of the house, calling out a cheery greeting.

A middle-aged woman answered to the door. Without saying a word, the woman took my empty bottles and disappeared into the house where I heard her talking to a man in Welsh.

A little boy came out to chat to me while I waited for the water, but he spoke no English, and I spoke no Welsh, so conversation wasn't much of a two-way thing. Looking for relief from the awkwardness, I turned to the dog and chatted to him instead. For all his barking and growling, he was actually a rather jovial creature, and conversation was satisfactory.

After ten minutes or so the woman re-emerged with my bottles in hand. I wondered what had taken so long, but I thanked her profusely and returned to Vlad where we enjoyed a long, cool drink before continuing on our way.

There were gates at regular intervals along the track and I decided it was high time Dakota learned the fine art of manoeuvring them with me on board. It was a useful skill to have and would save me from either having to dismount every time we came to one, or else waiting for five minutes while Vlad tried to manoeuvre them with Oisín.

Communication and co-ordination between the two of them was still a work in progress.

Dakota took to his lesson with timid willingness – trying his best not to panic at the swinging metal beside him and the sound of bars scraping over the rocks as I dragged or pushed the gates open to let him pass, then turned him around to close them once more. After the first few, he'd soon got the hang of it and became almost enthusiastic about the manoeuvre. At one gateway, however, Dakota didn't wait long enough before trying to squeeze through the opening. My stirrup caught in the bolt of the half-open gate and my knee banged hard against the gatepost. Dakota then panicked and bolted, tearing the stirrup from the saddle and crushing my knee as he went. When I'd managed to bring him to a halt and calm him down again, I dismounted and led him the rest of the way. Luckily the only damage was one ripped stirrup leather, a torn strap on my saddlebags, and a badly bruised knee.

16

A Mishap in the Mountains

'Oisín!' I cried, panic-stricken, watching helplessly as my horse tumbled backwards down the steep bank and landed on the hard rocks of the shallow stream behind him – right on top of Vlad.

Vlad will probably never forgive me for the fact that in that awful, heart-stopping moment, my first thought was for the welfare of my beloved horse, and not for him.

We were in the Brecon Beacons, half way up a mountain and several miles away from the Dan Yr Ogof caves and Shire Horse Centre where we'd stopped the night before.

Heavy clouds had obscured the mountains and a fine rain was falling when we awoke that morning. We'd enjoyed a hearty breakfast in the on-site café before we set off, following a steep track up into the mountains. After a hundred yards or so, we lost sight of the valley below as the cold, wet cloud enveloped us on all sides. We could barely see more than two hundred yards ahead as we picked our way between the rocky outcrops that loomed, dark and eerie through the fog. A smattering of small stones lay about us, sticking vertically out of the ground like strange, grey fingers – or as though they grew there like plants, with roots embedded deep in the mountainside. All was

A Mishap in the Mountains

silent here, save for the mournful bleating of lost lambs searching for their mothers.

We wandered through the trailing mist that rolled and tumbled down the rock-strewn peaks around us, lifting slightly with a gust of cold wind, before closing in on us once more. I let my mind drift, becoming happily lost in daydreams. Dakota strode ahead of our fellowship, picking his sure-footed way amongst the rocks of the well-worn track as it wound through treacherous bogs and beside deep pools of murky, rush-encircled water. It was easy to picture epic scenes from the Mabinogion or Arthurian legends playing out in this bleak, mist-lagged landscape that could be home to any number of giants, nameless monsters, or gateways to another world.

After a while we came to a stream and the path dropped sharply down a steep slope to stepping-stones below. There was no way we could get the horses down, so we looked to the left where the stream tumbled away across a bog, and then to our right up the bare hillside, searching for a way to get safely across.

There was a good, shallow crossing place just up from where we stood but Dakota flatly refused to step down into the water, and after a few minutes he spun round and leapt nimbly up the steep bank beside us.

I dismounted and tried to lead him across the river, but again, he was having none of it.

In the meantime Vlad was trying to encourage Oisín to cross the stream. It was narrow, shallow, and not very fast flowing. It should have been no problem at all ... but, taking a leaf out of Dakota's book, Oisín, too, was refusing to budge, and eventually he also spun and tried to leap up the bank.

Oisín is neither as agile, not as lightly built as Dakota, and I watched in horror as he lost his footing, overbalanced, and tumbled backwards into the water, landing on top of the still-mounted Vlad.

The whole thing happened in slow motion as a series of panicked thoughts raced through my head. What would we do out here in the middle of nowhere if Oisín got seriously injured? How would we get help? How would we get him off the mountain if he was hurt? What if he broke a leg and had to be put down? I felt the panic rising as all the worst possible scenarios raced through my head. Oh God! Please no!

I held my breath, aghast, as Oisín scrambled to his feet, climbed out of the stream, shook himself, and then began to graze. He was unscathed.

Phew. I exhaled. Then I turned my attention to Vlad.

He was ok – apart from a slightly sore ankle from where my eight hundred-odd kilo draught horse had landed on top of him. He'd live, I decided harshly, and turned my attention back to Oisín to make sure that he really was OK.

After righting himself in the stream, Oisín had had the good grace to get out on the far bank so Dakota now splashed placidly through the water to be with his friend.

Vlad's ankle was so sore that he was hopping, barely able to walk, and he needed help to get back on Oisín before we set off across the mountains once more.

The track here was overrun with walkers and strings of tired, wet Duke of Edinburgh Award students who looked miserable and fed up. Dakota decided he didn't like the walkers with their brightly coloured raincoats and large rucksacks full of gear, and so he reverted to being a nervous wreck – tense and ready to charge off at the first sign of trouble. My happy daydreams of earlier in the day scarpered, my mood plummeted, and I was keen to get far away from these dangerous hills with their horrid bogs and treacherous streams!

When we finally made it off the mountains we found that below the all-enveloping cloud the weather was actually pleasant and warm. We soon dried out, Dakota relaxed again, and the rest of the day was uneventful as we sped along the quiet country lanes towards our stop for the night, several miles west of Llandovery.

Our hosts – Jacqui and Phil – gave us a warm welcome, with plenty of much needed beer, a good curry, some more beer, and then copious amounts of gin and tonic which soothed our frayed nerves and dulled the pain in Vlad's ankle. Before we knew it we were rather tipsy, and by the time we staggered back across the fields to the caravan that was to be our home for the night, we realised we were actually quite drunk!

A Mishap in the Mountains

When we surfaced in the morning we both felt so rough, and Vlad's ankle was so sore, that we decided to have a day off. Actually, it wasn't until three days later that we finally left our lovely hosts and hit the road again, feeling well rested and refreshed – the traumas of that awful day in the Beacons now a distant memory, erased by plenty of good food, good company, and large quantities of alcohol.

Several months later when Vlad's ankle was still hurting him we realised he must have broken it in the fall – and I've never heard the end of it!

17

Anything is Better than a Tent

'Do you need somewhere to stay for the night?' A man in red van pulled up alongside us as we made our way out of the sleepy little village of Llanpumsaint. We'd barely covered nine miles since our last stop where we'd spent a delightful evening with Vic and Rog, two chefs who had tired of the high life jet-setting around the world to cook for the rich and elite, and had put down roots in a hill-farm on the edge of Brechfa Forest.

It was a bit early in the day to think about stopping, but people going out of their way to offer us somewhere to camp was a rare occurrence and couldn't really be refused.

'It's just up here on the right,' the man said in a thick Welsh accent, and proceeded to rattle off directions to his property. 'I won't be back till later, but make yourselves at home!' he said, then drove off up the road.

We found the property, stuck the horses in the field beside the house, and then made camp in the enormous barn where our host had said we could sleep.

I don't particularly like camping, so my theory when travelling is that anything is better than a tent. This philosophy hasn't always served me well however, and as a result I've spent a few uncomfortable nights in mouse-infested caravans and wet barns full of sheep and dog excrement. This barn, however, was a nice clean one, which our host used as a workshop. It had a well-swept concrete floor, and there was even a toilet we could use. This was the height of luxury!

One of the best things about life on the road is that you learn to really appreciate the little things, and even a barn can seem as good as a five-star hotel if you thought you'd be sleeping in a tent. I suppose the secret is that if you have expectations, you will nearly always be disappointed; but if you expect nothing, then everything you are offered is an absolute delight.

The man returned home later that evening and invited us in for drinks. At his request, I brought my fiddle and played some tunes, sipping gingerly at the enormous glass of wine that had been thrust into my hand.

Our host was in his fifties, with greying hair, and he owned a large electrical company. Although successful in his career, he seemed unhappy, and tired of the daily slog.

'I'd like to do what you're doing,' he said wistfully. 'Give it all up, leave my troubles behind, and ride off into the sunset.'

'It's nowhere near as romantic as it sounds!' I said, laughing, and Vlad agreed. Undeterred, our host began to press for more information.

'How do you plan your routes?' he asked Vlad. 'And how many miles do you ride each day?'

Vlad shrugged. 'You'd better ask Cathleen. She does all the navigating; I just follow her directions!'

As a relatively inexperienced rider, Vlad had enough on his plate trying to manage Oisín and Spirit and come to terms with life on the road. He didn't need to add the stress of route planning and navigation to that. Besides, his sense of direction was appalling so we'd probably have ended up in France if he'd tried!

I answered our host's questions, telling him about the issues we'd found with bridleways, and how we tried to avoid main roads and towns, sticking mostly to quiet back lanes.

'How do you find places to stop?' our host then asked, again directing his question to Vlad.

Fiddler on the Hoof

'Mostly we've been staying with friends of Cathleen – people she's stayed with on her other journeys,' Vlad said. 'She's done several long distance rides by herself in the past. This is my first time, so I'm still learning. She knows more about travelling on horseback than I do.'

So again I answered our host and told him how we went about finding places to stay.

'What do you do when your horses need new shoes?' our host wanted to know next – still asking Vlad, who once again politely redirected the question to me – because along with all the other logistical stuff, I dealt with finding farriers, too; and once again I provided him with the answer.

'Where did you get your horses from?' he asked Vlad. This was really starting to annoy me now.

'They're Cathleen's horses,' Vlad replied. 'She rescued them.'

'How would I find a good horse for such a journey?' he queried.

'I'm new to horses. You'd really have to ask Cathleen,' Vlad said again.

But for whatever reason our host wouldn't ask me anything about the logistics and technicalities of life on the road, and continued to direct all his questions to Vlad.

Eventually – tired of being ignored and fed up with this irksome triangular discussion – I sat back in my corner and made pretty background music while the men talked.

As the daylight faded and darkness fell I set down my fiddle and tuned back into the conversation.

'I'm lonely up here all by myself,' our host was saying, as he started on the second bottle of wine. 'I need a good woman – someone to cook for me, and do my cleaning. I don't suppose you know of anyone?'

Vlad shook his head. He did not.

'I had a good woman – my girlfriend – but she left me,' he said forlornly. 'She was beautiful – always had her hair and make-up done immaculate, like. She dressed herself nicely, too – really made an effort. I like that in a woman. She looked after herself, you know? And she was a good housekeeper.' He sighed and took a long swig from his glass. 'I was married once,' he continued with the same melancholic nostalgia. 'Thirty-seven years!'

'What happened?' Vlad asked.

'She divorced me when the kids left home,' he replied.

Somehow this news did not surprise me.

'I need another woman, one who will do my cooking and cleaning – a live-in housekeeper to share my bed,' he repeated again, and I cringed.

It struck me that just because someone was kind enough to offer us hospitality and a place to stop for the night, didn't mean I actually had to like them; and so far, I wasn't seeing a whole lot to like about our host. At best, I pitied this man: he may have been materially rich, but his out-dated attitudes to women and traditional gender roles, in my view, rendered him incredibly poor; but at worst, he was downright shallow and misogynistic.

When our host offered us a room in the house for the night, to Vlad's surprise, I declined and said the barn would be just fine, but thanks anyway. Although anything is better than a tent, there are times when a barn is preferable to a bed.

18

A Chance Encounter

The following morning, as we tacked up the horses, I decided I was sick of our saddles, and if we wanted to continue the journey then we would need new ones.

We had set off using treeless saddles – not because I particularly like them, but because that was all we had. We soon discovered that treeless saddles were no good for long journeys such as ours because they don't distribute the riders' weight properly, no matter how many shims and special pads you use. They cause lots of localised pressure and they slip horribly with packs attached. After more than a month of covering twenty-odd miles a day, five or six days a week, we'd noticed that the horses were sore behind their shoulders where the pommel sat, under the stirrup bars, and under our seat bones. Oisín was already starting to show white hairs from the pressure, and that sore under the saddle had only just healed.

Something had to be done … but what? How were we going to source and fit two new saddles here in the middle of nowhere in Wales? And would getting new saddles really solve our problems, or would they just create new ones? I was in low spirits with the anxiety of it as we set off that day.

I am by no means a religious person, but over the years I have developed a strong belief that if you want something badly enough,

A Chance Encounter

then Fate, Fortune, Providence, the Universe – whatever you want to call it – will almost always provide. That belief has developed over the years due to countless examples of the right things appearing at just the right time when they were most needed – like John the farrier, or the vet at Emma Bowyer's, to name just two examples from the journey so far.

To my great relief it was no different with the saddles.

On the very day that I was beginning to despair and wondering how, in good conscience, we might continue the journey, that unfailing benevolent Universal force did what it usually does and proffered a solution to our problem. That solution came in the form of Alan and Judith Sterling.

We were making our way along a narrow, single-track lane in the middle of nowhere heading towards Cynwyl Elfed when I heard a vehicle approaching behind us. I pulled onto the wide verge and let Dakota drop his head to graze for a minute until it had passed. I was walking in front with Dakota that morning, and Vlad, Spirit, and Oisín were a good hundred yards behind us, wrestling with Vlad's ever-slipping saddle packs. After a few minutes, the vehicle still hadn't passed us. Looking back, I saw Vlad talking to the driver of a Land Rover, and after a few minutes I wandered back to see what was going on.

'They're saddle fitters!' Vlad said with excitement, handing me a card that read: 'Specialized Saddles – the saddle with an adjustable fit'.

A bell rang in the fog of my distant memory. I had heard this name before … but where? I wracked my brains, and after a few minutes the penny finally dropped. I had been discussing equipment with fellow Long Rider Vyv Wood-Gee on my way down from Scotland the year before and she'd shown me her amazing, lightweight, fully adjustable endurance saddle. It was a Specialized Saddle, she told me, and the only people in the UK who sold them were based in Wales. I'd even made a note of the name, determined to look them up when I got home, but my subsequent searches for 'Specialised Saddles' had rendered nothing. Now here we were, on a tiny road in the back of beyond, on the same day that we had reached our wits' ends about the saddles, talking to those same people.

Alan and Judith introduced themselves. They wanted to know where we were going, what we were doing – and of course, they wanted to know what equipment we were using.

We told them of our issues with the treeless saddles: the pressure, the sores, and the constant slipping. They confirmed what we already knew – treeless saddles are no good for travelling long distances and carrying equipment; they cause too much damage to the horse.

Now don't get me wrong, I'm not against treeless saddles. In the right circumstances I'm sure they work well for both horse and rider – but they definitely weren't working for us on a long ride, covering an average of a hundred miles a week!

People in the horse world can get quite worked up about certain things like shoes versus barefoot; bitted versus bitless bridles; treed versus treeless saddles. In my experience of these matters equestrians tend to pick their side of the fence and then stick doggedly to it, regardless of whether it's working for their horse or not. They become like religious fanatics, refusing to acknowledge anything that goes against their beliefs and denouncing anyone who disagrees with them. Unfortunately in these instances it is only ever the horse who suffers. If you ask me, the true mark of a good horseperson is adaptability, an open mind, and a willingness to put aside their own ego and ideologies in order to do what is best for the horse – whether the rest of the world agrees with it or not.

After a brief discussion, Alan asked where we were headed and offered to come out and meet us at our next stop to see if he could help – either by modifying our existing saddles, or by fitting the horses with some new ones. It was an offer we couldn't refuse.

Still marvelling at the perfect timing of the whole thing, and feeling rather elated with our faith in Providence and the Universe at an all-time high, we covered the twenty-four miles to our next stop with ease, as though it were little more than a saunter around the block. Even the thunder, lightning, and heavy rain that set in late in the afternoon and saw us soaked to the bone did nothing to dampen our high spirits.

'What were you doing on that road?' I asked Judith when she and Alan turned up a few days later in the still pouring rain with two prototype Specialized Saddles, which they began fitting to our horses. I was still in a state of happy shock at how perfectly timed our encounter had been.

'We were following the sat-nav,' she replied. 'We thought it was a bit odd when it routed us down that tiny road and debated ignoring it because it would have been easier to take the main road – but then we thought "Why not?" because we'd never been that way.'

I laughed, shaking my head. You couldn't have made it up! If that wasn't an example of perfect synchronicity, and the Universe conspiring in our favour, then I didn't know what was!

The Specialized Saddles work on a system where the wide panels underneath the saddle – which rest either side of the horse's spine – are covered with a Velcro strip, to which shims of high-density foam can be attached. The different shaped shims and wedges allow the saddle to be fitted perfectly to the horse, and moreover they allow the saddle-fit to be adjusted as the horse builds muscle and changes shape. The system is a marvellous and innovative one.

Alan worked fast, expertly adding shims to create an even contact under the saddle along the horses' backs, with plenty of spinal clearance. Within an hour not only did our horses have two well-fitting saddles, but Alan also threw in some extra shims for us to take with us – just in case we needed to adjust anything later on in the journey.

We gave Alan and Judith our heart-felt thanks when they left. Had it not been for them, our journey would have ended there in Wales. What an extraordinary stroke of luck – if you can call it such – that chance encounter had been!

19

Plas Dwbl

On the day of our amazing encounter with Alan and Judith Stirling, we had arrived at Coleg Plas Dwbl at the foot of the Preseli Mountains in Pembrokeshire. Plas Dwbl was a Ruskin Mill college for young people with special needs, and a sister-site to Vale Head Farm where I'd spent a happy evening on my journey the year before.

The college ran arts and crafts workshops for teenagers and young adults with a range of learning difficulties and autistic spectrum conditions. As well as the workshops there was a large farm and vegetable garden, both of which were run on biodynamic principles.

Laura, the head farmer, was an old acquaintance of mine. She'd had a farm just up the road from where I lived in Cornwall before she moved to Wales to take a job at the college. When she heard about our journey she invited us to stop off at Plas Dwbl on our way through. There was plenty of land for the horses, an empty flat we could use, and we were welcome to stay as long as we liked. Most of the students were on their summer holidays so we'd not be an inconvenience, Laura assured me.

The offer was a welcome one. When you're travelling on horseback time loses all meaning and importance. The days, dates, and hours fall away into nothing, replaced by gradually shifting landscapes, changing points on the vast horizon, and the steady clip-clop of horses' feet under the boundless skies. A single day can seem like a week, and a week as long as a lifetime because you become so wrapped up in the moment. Thinking ahead beyond finding a place to stop for the night becomes a wholly futile exercise, and organising anything more than a day or two in advance is almost impossible. If ever there were a good way to escape our fast-paced modern world and immerse oneself wholly in the present, then horseback travel is it! Vlad and I had been so caught up in the day-to-day challenges of the journey that West Wales and the end of the UK part of the adventure had snuck up and caught us unawares. We hadn't yet organised any transport to get us over the sea to Ireland. In fact we'd barely given it much thought at all! A few days to rest and plan the next leg of the journey would be very welcome indeed.

We arrived late on a Saturday afternoon – cold and wet after a good soaking in the earlier downpours. Laura met us, showed us a nice field for the horses, and then took us up to the flat where there was a comfortable bed, a fully functioning kitchen, and a shower. There was even a washing machine we could use!

'Help yourself to any vegetables from the garden,' Laura said. I'd already been eyeing up the poly-tunnels full of ripe onions, potatoes, beans, tomatoes, lettuce, cucumbers, courgettes, and row upon row of delicious chard, kale, and Cavalo Nero. We didn't need to be told twice, and we feasted well during our stay.

We were just sorting out our wet gear, and hanging things up to dry in the barn, when a young man suddenly appeared on the scene. He was his early twenties, and had lank, greasy hair that reached his shoulders but was shaved short at the sides. His face was gaunt, with intense blue-green eyes, and a hooked nose that put me in mind of a buzzard. He was barefoot and wore a long woollen kaftan that was open at the sides and he didn't appear to be wearing anything underneath it. Laura introduced him as Rhion, the apprentice who was doing his two-year-long biodynamic agricultural training at the college.

Rhion lived alone on site. When all the students and their support workers had gone home at the end of the day, there wasn't a soul about the place and the nearest neighbours were a good half a mile away. It was a lonely existence out here in the middle of nowhere in the bleak landscape. Rhion entertained himself in the long hours of solitude with mediation, playing his tin whistle, reading books on spirituality, philosophy, and Celtic mythology, and by fasting for extended periods of time. He was almost at the end of a seven-day fast, he told us – which went a long way to explain the gaunt, hollow cheeks, and the rather spaced out expression. Although he said he wasn't lonely, he clearly missed human contact because he simply wouldn't top talking. He followed us into the flat and made himself comfortable on the sofa in the kitchen. While we prepared dinner he talked in a ceaseless monologue about a variety of strange and esoteric topics. He spoke seriously, and with great intensity. If Vlad or I wanted to get a word in edgeways then we had to talk loudly over his steady flow of words until he finally stopped to listen. It was quite exhausting after a long day's journey.

We felt guilty eating a feast of freshly harvested vegetables in front of Rhion while he fasted, but he seemed not to mind because he was glad of the company. He sat happily perched on the sofa – talking and talking – with his knees drawn up to his chin in such a way that my suspicions about his lack of undergarments were confirmed when I got an inadvertent eyeful of a whole lot more than I'd bargained for!

By midnight I was tired and unable to take any more of the heavy monologue, so I slipped away to bed, shamelessly leaving Vlad as a captive audience to Rhion's ramblings.

The following day, it was raining again. We'd noticed that the fields and verges appeared greener the further west we came. It was the first real green we'd seen in a long time. Clearly this part of Wales wasn't experiencing the same drought and infernal heat that had plagued the rest of the country for the best part of two months!

I spent most of our first day at the farm trying to organise transport to Ireland, and searching for somewhere to unload the horses once we got there. Vlad, on the other hand, spent the morning fixing a broken tractor – he's a dab hand at that sort of thing!

We were both successful in our endeavours: by lunch time the tractor was back up and running, and Nathan Deakin – a local equine transporter who was heading over to Ireland with an empty lorry at the end of the week – said he would be happy to take us across the sea. West Cork Equine Centre, near Bandon, had also kindly offered us a place to unload the horses and spend a night before starting out on the next part of the adventure. It had all fallen seamlessly into place. Now all that remained was for Vlad to make a decision. He had only agreed to ride with me for a few weeks to get a taste for equestrian travel and life on the road, but now that we were about to cross the sea, it was time for him to make a choice:

'My only commitment had been to go with Cathleen as far Pembroke, and I had the option of returning home while she continued alone.

'Those first weeks on the road had been the toughest time of my life, but the most amazing, too. After four hundred miles the walking no longer bothered me, as my soft skin soon hardened and my muscles became accustomed to many hours spent in the saddle. I was starting to understand Oisín's needs – and he mine. Our relationship was improving, and I felt I had slightly more control over this enormous stomach on legs.

'The journey so far had provided me with unique experiences that only the road can bring. The hospitality of the people we'd met had given me deeper insights into the better side of human nature and left me feeling overwhelmed. At the time of our travels I was reading The Alchemist – one of Paulo Coelho's best sellers – and I was looking for omens. Synchronistic events and encounters had become a recurring theme on our journey. Needless to say I felt I was on the right path and not only surviving, but growing and thriving, too. If England and Wales had provided so much, then what would Ireland have to offer, I wondered?

'I wasn't ready to go home just yet. I definitely wanted to continue.'

With everything sorted, we had nothing to do but relax, fill up on fresh vegetables, and test our new saddles as we explored the Preselis and surrounding bridleways.

The afternoon before we left, we took the horses out onto the hills one last time. It was exhilarating cantering through a sea of long, windswept grass; the early evening sunlight merged with dark clouds that billowed in from the west, casting a strange light over the land. Here and there circles of standing stones and ancient burial chambers lay scattered across the barren hillsides, and the rock-strewn tors looked like the scaly backs of sleeping dragons – illuminated by the golden storm-light against the backdrop of black sky. This place reminded me of Bodmin Moor back in Cornwall which we'd left behind us five long weeks and four hundred miles ago. It would be several months before we set foot on this island again.

Across the sea to the west lay Ireland – that magical land of myths and legends, folklore and fairy tales, and wild, enchanting music. There the journey would really begin.

Nathan Deakin arrived at Plas Dwbl at eleven o'clock in the morning of the 2nd of August. Tensions had been running high as Vlad and I were packing up and preparing to leave – unsure of what to expect from Ireland, and worried lest the horses refuse to load. Although both horses were now experienced travellers, neither had spent much time in a lorry and I doubted whether either would be keen to adopt this new mode of transport.

As anticipated, both Dakota and Oisín flatly refused to board the lorry, and Spirit voiced her concerns at the whole proceedings with loud howls and whines which didn't help matters in the slightest!

With lots of coaxing and a little food, Oisín was eventually persuaded up the ramp but immediately objected to being shut in a partition without his friend, and so began rearing and pawing at the sides of the lorry. Poor Nathan looked more than a little worried – and who could blame him? A chunky draught horse can do a lot of damage to a lorry! In the end – because Dakota was digging his heels in and exercising his remarkable talent for moving backwards as quickly as he can move forwards – we dug out some long ropes and, avoiding his flying hind feet, we finally got him into the lorry. Oisín immediately stopped panicking, Spirit stopped her cacophony, and at long last we were ready to go.

Part 2: Ireland

The route across Ireland

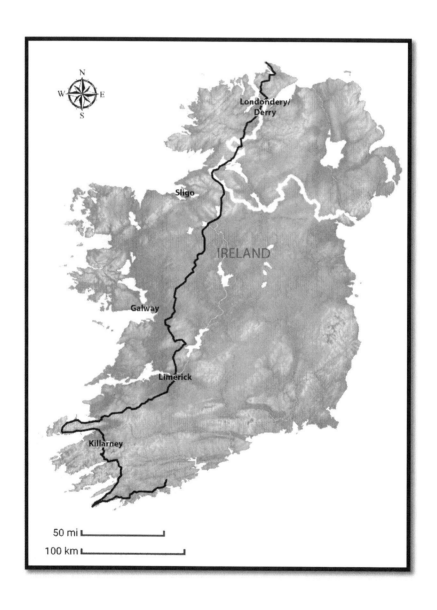

20

First Impressions of Ireland

It is a strange feeling arriving in a country where everything is different, just not in an obvious way. The language is the same, the culture not too far removed from one's own, yet things are somehow ... well, different! And the first major difference we noticed in Ireland were the roads.

In Britain, it is customary for drivers to slow down when approaching a horse and to give it a wide berth. In Ireland, this was certainly not the case! Motorists seemed to have absolutely no idea how to drive sensibly around horses, and appeared oblivious to the fact that a) a horse can spook at almost anything, and b) a spooked horse can cause a lot of damage to itself, its rider, a car, and anyone in it. In fact, many of the equestrians we met in Ireland said they would never ride on the roads because it was simply too dangerous, and so they rode in sand-schools or fields instead.

I was thankful that our horses were both sensible in traffic, and that we were kitted out with brightly coloured packs, hi-viz leg bands for the horses, and hi-viz vests for ourselves, along with luminous coloured jumpers – all in a startling array of yellows, greens, oranges and pinks. To say we were visible from a mile off is probably an understatement, yet it did little to deter the speeding maniacs on the Irish roads. On our first day in Ireland, a lorry attempted to overtake us at speed, giving us no space at all. At exactly the same moment, a curious herd of cows in the field beside the road came charging full pelt in our direction, startling poor Dakota who shot out into the road, and only narrowly escaped a collision with the lorry. Thankfully the lorry driver had the good sense to stop and wait until I had calmed Dakota down before he proceeded to overtake. From that moment on I adopted Vlad's more pig-headed approach to traffic management and rode in the middle of the lane so the traffic would have to slow down and think twice before squeezing past.

I think it is fair to say that most of the terrible driving we encountered was not born of malice, but rather from sheer ignorance, not helped by the fact that so few equestrians ventured out onto the lanes. We also quickly discovered that, on the rare occasion when a driver would pass sensibly, if we put a hand up to thank them for the courtesy, they would beep their horn in a cheery greeting right next to the horses, which would send them flying off down the road in fright! After a few days, however, Oisín and Dakota became desensitised to both the lunatic drivers and the car horns, and they soon learned to ignore the free-ranging dogs which ambushed us in every gateway we passed. As for getting off-road? Well you can forget it! There is one bridleway in the whole of Ireland, which is located in the Miskish Mountains at the bottom of the Beara Peninsula in County Cork. It runs for a grand total of 17 kilometres, and in order to reach it you'd have to ride for many miles along busy roads. There are but a scant handful of footpaths across the whole of the island, too. It seemed that the concept of public access to the land was an almost non-existent one here, and I suddenly found a new appreciation for the bridleways back home.

On the first day we left West Cork Equine Centre and headed south to Timoleague, where the ruins of a large 13th century Franciscan abbey stood overlooking the estuary at the mouth of the river Arigideen. From there we picked our way along quiet roads to Clonakilty and, opting to avoid the main road and all its heavy traffic, we rode right through the middle of town. The houses and shop fronts were painted in pretty shades of blue, pink, red, green and yellow, and colourful flowers grew thick in the many window boxes and hanging baskets, giving the town a bright, cheerful feel. People stopped in their tracks to stare at the strange sight as the horses clattered their way up the main street, before we turned off on a side road to pick up quiet roads again.

One lane we came to on the outskirts of Clonakilty had clearly been abandoned by motorised vehicles some while ago because it was covered in a thick layer of soft, green turf. The road still bore an optimistic speed restriction of 80kmph. We did our best to break the speed limit and cantered happily along it for a few miles until we hit tarmac once more.

After a pleasant day's ride, we finally arrived at Tullineaskey Equestrian Centre where Vlad, Spirit, and I spent a comfortable night in the gallery overlooking the large indoor sand-school while the horses grazed in a paddock out the back.

In the morning we headed up into the hills to avoid the busy main road. Grass grew thick in the middle of narrow lanes, wild fuchsia adorned the hedges, and we hardly met any traffic at all. The road wound between small undulating hills covered in pasture, gorse scrub, and patches of bog. Little farms lay scattered throughout the countryside, and everywhere we looked there seemed to be a large, new-build house plonked right in the middle of each available field. There was no definitive point where one village ended and another one began. It felt as though we were riding through a continuous built-up area in the middle of nowhere, which gave the landscape a rather cluttered feel.

As we approached Skibbereen we decided to look for somewhere to stop for the night. Although we passed many houses, we saw no-one to ask. Luckily for us, our host that morning had given us a phone number for one Jane Scully.

'Give her a ring if ye get stuck. I'm sure she'll put ye's up. She's good craic!' we'd been assured.

So we rang Jane and she didn't hesitate in offering us a place to stay.

21

The Baltimore Regatta

Jane Scully lived on a farm between Baltimore and Skibbereen. She had a heavy Cork accent, looked to be in her late forties, and her thick, shoulder-length brown hair – which had a copper tinge from days spent outside in the sun – was streaked through with strands of silver. She came from a family of horsemen, bred top quality event horses, and what she didn't know about equines probably wasn't worth knowing.

Jane was sharp, could read a person a mile off, and she didn't mince her words. She was the sort of person who spoke her mind, took no nonsense from anybody, and I got the distinct impression that she preferred the company of her dogs and horses to that of people. Although life had dealt her a few knocks over the years, under her slightly hardened exterior she was warm and empathic, and she had a heart of gold.

The weather was nice and the evening warm, so after making camp and settling in we sat out in front of Jane's house drinking wine and chatting. As the evening drew on, I fetched my fiddle and played a few tunes.

I was seven years old when I started learning the violin. My mother had been determined that both my sister and I would pick an instrument and stick to it. It was a good skill to have in life, she'd insisted, whenever we complained. So my sister had chosen the harp, and I had chosen the violin.

We both learnt classically, but I soon decided I wasn't keen on classical music. It was too rigid, too structured, and it lacked the raw energy of the traditional Irish tunes my father was always listening to. I wanted to learn to play music like that!

'You can learn that in your own time,' my violin teacher told me. 'It's easy. Just find the sheet music and you'll be able to play it.'

Except when I found the sheet music for the tunes I loved, and diligently played the notes that were in front of me, it sounded nothing like the recordings I was listening to. In fact, no two recordings of a tune even sounded the same! I was definitely missing something here. Disheartened, I gave up trying and shortly after I ditched the violin altogether in favour of horses.

I was nearly twenty-two when a friend and I went to see the Peatbog Faeries in St Ives. They were a high-energy Scottish band playing a mix of original and traditional Celtic tunes with fiddle, whistle, and bagpipe melodies set to heavy electronic dance rhythms and plenty of bass.

I was enthralled. Here was the kind of music that sent my heart leaping into my throat, and put my feet moving almost against my will. I stood in the crowd watching the fiddle player's fingers flying over the strings, perfectly synchronised with the swift, smooth movement of his bow to produce the most enchanting, soul-capturing sound – and there wasn't a sheet of music to be seen. Why couldn't I play like this? I should be able to play like this!

Exasperated, I went home that night, dusted off my violin and, with a renewed determination, I set about learning traditional Irish music – not by following the musical notation this time, but by listening to the melodies, feeling the music, and letting my heart and my fingers work together to give my own rendition of a timeless tune that had been passed down through the centuries from one musician to the next, each making it his own before passing it on again. This was how music should be played!

Fiddler on the Hoof

'It's the Baltimore Regatta this weekend,' Jane said as the last note faded and I packed my violin away. 'They'll probably have music on. Why don't you stop here for a few days? You might pick up some tunes.'

We leapt at the chance, accepted her offer, and the next day we headed into Baltimore with my fiddle in hand.

The sun glinted on the green-blue water of the harbour through a forest of tall masts. The bay was filled with sailing yachts and motorboats, and hordes of people dressed in designer sailing gear and life jackets were milling about on the seafront. All the pubs in the little town were heaving, alcohol was flowing, and on the small terrace in front of a French restaurant, a four-piece band were playing a mix of blues and rock covers. There was a guitar, a banjo, a fiddle, and a bass – but no traditional music. My heart sank.

'Go on, get up there a play a tune with the band!' Jane pushed.

'I can't just go and hi-jack their set!' I said, shocked. 'And besides, they're not even playing traditional music.'

'Leave it with me,' Jane said, and vanished into the crowd. A little while later, when the band were on a break, Jane reappeared with the banjo player in tow and introduced him as Brendan McCarthy.

'These pair are riding their horses around Ireland raising money to help disabled people,' Jane explained, matter-of-factly. 'Cathleen's a fiddle player. She's very good.'

'Come on up and play a tune with us,' Brendan said. 'Which ones do ye know?'

So I joined the band for a few traditional tunes while Jane whipped round the crowd collecting donations.

'Dig deep! Dig deep! It's nice to be nice,' Jane chivvied. 'Ah come on now, it's for a good cause helping those less fortunate than yourselves!'

Jane was not someone I would dare to refuse – a sentiment that seemed to be shared by most of the people there, so the donations rolled generously in.

At my request, the band then played a couple of local reels – The Banks of the Ilen, named after the river that runs through Skibbereen and meets the sea at Baltimore, and a reel named after Tom Billy Murphy.

Tom Billy Murphy was a blind musician from the Sliabh Luachra – a region that borders Counties Cork, Kerry, and Limerick, famous for its culture and for giving rise to many of Ireland's most renowned musicians and poets. Tom Billy was born in the late 18th century and, at the age of thirteen, contracted polio which left him blind and lame in one leg. Fortunately, he'd shown an aptitude for music and soon became a celebrated musician and fiddle teacher from which he could earn himself a decent living.

Tom Billy's preferred mode of transport was a donkey. It is said that the faithful animal only needed to be shown a destination a couple of times before it knew the way, and so would reliably carry its master on his rounds from the house of one student to the next, or between the public houses where he played. Many traditional Irish tunes from the region bear his name.

We left crowded Baltimore late in the evening and drove into Skibbereen. Although a small town with a population of little over 2,500, it seemed that every other brightly painted building in the town centre was a pub.

'There aren't as many as there were,' Jane said when I commented on the fact. 'There used to be forty-seven pubs, but now there are only twenty-six.'

There was live music at The Corner Bar. Two of the musicians from earlier – Brendan McCarthy and one of the others, whose name I didn't know – were there entertaining a small crowd of locals. This time they were singing old ballads and Irish rebel songs. It was rousing music filled with longing for freedom from the oppressive Brits who had occupied these lands for centuries, tales of heroic deeds performed in the many rebellious uprisings, and laments for those who had lost their lives fighting for Ireland's independence. Everyone in the bar knew the words and joined in on the choruses. Guinness flowed, music played, and it was very late when we finally tumbled back into our tent – exhausted but content, our heads reeling from all the Guinness and good music.

Fiddler on the Hoof

We awoke with throbbing hangovers in the morning but, determined to hit the road again, we went to fetch the horses from the enormous field where they had spent the last few days gorging on rich grass. There is something about the grass in Ireland – it has a certain quality of green to it that I've never seen anywhere else in the world. I'm not talking about the dark green, chemically fertilised grass that surrounds dairy farms and gives horses colic, I mean the natural grass that grows on the open hills and verges of the aptly-named Emerald Isle.

Evidently the grass possessed a certain quality of goodness, too, because – for the first time on the entire trip – both of the horses flatly refused to be caught and galloped madly around the field, kicking their heels to the sky and leaping about in delight. We had to resort to rattling a bucket of feed in order to catch them and to my dismay I noticed that Dakota was missing a shoe!

'You'll be lucky to find a farrier this week,' Jane said pessimistically. 'They'll all be at the Dublin Horse Show.'

I called the number she gave me, anyway, and luckily Billy – one of the three farriers for the whole of West Cork – was not in Dublin and agreed to come out to us.

He was a quiet man, with a dry sense of humour, and an accent so thick that I had to wait a few seconds after he spoke to decipher what he had just said. He was a good farrier though, and Dakota behaved perfectly while he pulled off the three remaining shoes – which were all very worn – and swiftly nailed a new set in place.

In Ireland they cold-shoe horses, which saves a lot of time and effort, and also a lot of money, so it wasn't long before we were able to load all our gear up, bid a fond farewell to Jane, and be on be our way. We were heading for the Mizen Peninsula.

22

Mizen Head

A baby cried, its shrill screams echoing around the walls of the stone tower and across the whole of Tory Island.

Balor, king of the Fomorians, contemplated the small, helpless creature that was clutched tightly against his daughter's breast.

'The child must die,' Balor's druid whispered in the giant's ear. 'If he lives, he will kill you.'

Balor looked at the child again with his one good eye, and beneath its seven coverings his poisonous eye burned. With that terrible eye he could wreak havoc and destruction on the land and his reign of terror had been a long one.

'Kill him,' he said harshly, unmoved by his daughter's heart-rending cries as she pleaded for her newborn son to be spared. The infant was taken to the window of the tower, where its mother was held a prisoner, and thrown into the dark waters of the Atlantic far below.

Balor breathed a sigh of relief. His fate had been averted; the prophecy would not be fulfilled.

But unbeknownst to Balor, the child did not drown.

Manannán, god of the sea, caught the baby – cradling him in the rolling waves – and brought him safely to shore. He named the boy Lugh, and raised him amongst the Tuatha Dé Dannan as his own son.

Fiddler on the Hoof

Mizen Head, on the Mizen peninsula, is the southern-most point in Ireland. Or at least that's what everyone believed, until they discovered that nearby Brow Head is actually a whole nine metres further south. Now Mizen Head's claim to fame is that it is the most South Westerly point in Ireland. And, since Mizen Head has traditionally been to Ireland what Land's End is to Britain – a good start or finishing point for an 'end-to-end' journey – that's where we decided to head.

The day was bright and sunny as we set off from our stop near Schull, where Linda and Jimmy O'Donovan – some friends of Jane – had kindly offered the horses a field for the night, and their lovely neighbours, Denis and Mary, had let me and Vlad sleep in an outbuilding which they used as a games room. We made our way down into the small village past a row of flapping flags. Dakota felt those were a serious threat to his safety and well-being, and so he bolted off down the road with me several times before I could eventually persuade him to pass them – which he did on tip-toes, with great trepidation and much snorting. We abandoned the main road in favour of narrow lanes that wound up into the quiet hills, past fields full of cows and donkeys. There seemed to be an awful lot of donkeys in Ireland and when we passed them, they would stand solemnly watching our progress with deep, thoughtful expressions on their wise old faces. Every now and then one would let out a mournful bray, which would send both the horses flying down the road in panic because neither had had any dealings with their long-eared cousins. Linda, our host from the night before, had said that female donkeys were much sought after by farmers in Ireland because they were widely believed to prevent 'redwater' in cattle – a tick-borne parasitic disease otherwise known as Bovine Babesiosis.

We came down from the hills and picked our way along the coast road. Rocky outcrops and small coves with beaches of soft, white sand stretched away to the sea, which lay calm and sparkling in the warm August sunshine. Were it not for a fierce wind that blew in from the Atlantic, it could have been the Caribbean!

Houses became sparser here and the hills more barren, and finally I felt a sense of space and empty, uninhabited landscape. After a few miles, as we neared Barleycove, the road split. One road led to Three

Castle Head, the other to Mizen. There was a man leaning on a gate in front of an old school as we approached the junction. He had long grey hair, a white beard, and a blue sailor's cap perched on his head. He asked us where we were going, what we were doing, and then filled us in on a little bit of local history.

'Over there,' he said, pointing to the opposite side of the headland, 'Is Three Castle Head. It is called that because there were two castles there.' He paused, and Vlad and I exchanged a confused look. This must be some sort of Irish logic. 'Only the ruins of one castle remain,' he continued, and our confusion only deepened. 'The castle has three towers, which give the headland its name,' he finished at last – as though the whole thing now made perfect sense.

As with all self-respecting castles, Dun Lough came with a ghost and some gruesome tales. A White Lady was said to wander the shores of the lake and death would come to anyone unfortunate enough to see her; there was a tower that dripped blood once a day to commemorate the last family to inhabit the castle – all of whom met with a tragic end, either by murder or suicide; and to top it off, there was a pile of cursed treasure that would bring great misfortune to anyone who found it. Luckily it was at the bottom of the lake – which was reputed to be bottomless – so chances of finding it were slim.

Enchanting as it sounded, we decided to give the castle a miss and rode on towards Mizen Head.

Mizen Head in Irish is called Carn Uí Néid, meaning the Cairn of Néid's grandson. In Irish mythology, Néid was the god of war, and his grandson was Balor, king of the Fomorians – a race of monsters and giants who terrorised the whole of Ireland. Balor had an eye so poisonous that when uncovered, it could wreak destruction on the world in the form of drought, blight, and the scorching summer sun. Legends say that the Tuatha Dé Dannan – a race of gods and supernatural beings – grew tired of being oppressed by the Fomorians and rose up to fight them, led by Balor's own grandson, Lugh.

The battle raged, blood flowed, and both Fomorians and Tuatha Dé Dannan alike lost their lives. Victory on either side was looking uncertain, until Balor entered the fray. Slowly, one by one, he began to remove the coverings from his terrible eye. With the first covering the bracken began to wither. With the second the grass shrivelled and dried. As the third one was cast aside, the land began to grow hot. With the fourth, smoke rose from the surrounding forests. With the fifth, everything grew red; and with the sixth it sparked into flame. As the seventh covering fell to the ground, unveiling Balor's poisonous eye, the whole countryside was set ablaze.

The Tuatha Dé Dannan retreated before Balor, unable to withstand the giant's fiery fury. Then, in a last, final attempt, Lugh cast his spear, piercing Balor's eye, and pushing it out through the back of his head where it destroyed the Fomorian armies behind him, reducing them to ashes. Balor fell to the ground, where Lugh beheaded the giant and the battle was won.

Some say this battle took place in County Sligo, and where Balor fell his eye burnt a deep hole in the rock that filled with water to become Loch Na Súl (Lake of the Eye); while others say he fell at Mizen Head, which is how it came by its name.

It wasn't hard to imagine an epic battle between gods and giants playing out in this wild and rugged landscape, but nothing remained save the name.

We followed the road past Barleycove, which was less of a village than it was a smattering of houses overlooking a long, sandy beach. From there the road became narrow as it climbed up around the steep, wind-swept side of the headland, where little thrived but gorse and heather. The land was divided into small fields by neat lines of stone walls in whose tenuous shelter only a few resilient wild flowers grew.

It was late afternoon when we finally arrived at the visitor's centre and café, which were just closing up for the day. A line with 'finish' written on one side, and 'start' on the other had been painted onto the tarmac by the café, so some obligatory photos were taken.

A kind lady from the visitor's centre came out and offered the animals some water. Then she helped us find a field for the night, about a mile back down the road, high on the windswept headland overlooking Barleycove and Browhead.

Mizen Head

The following day, as we headed back past Barleycove, an inquisitive Shetland pony climbed over a low wall and out into the road to get a closer look at our two horses.

Dakota took one look at the diminutive creature and decided this was not for him, spun sharply around, and made off up the road at a determined trot. Oisín – remembering that his ancestors were war horses who had carried knights into battle centuries before – arched his neck, puffed himself up to a size well beyond his meagre 15 hands, and tried to charge at the little upstart to run him into the ground.

Chaos ensued as I struggled to get Dakota under control, shouting instructions at Vlad on how to prevent bloody carnage – which he mostly ignored because he found the whole thing deeply amusing. I grabbed a passing Dutch tourist and told him how to catch the offending pony and stick it back in its field, which – after several confused looks and a minor scuffle – he did. The Shetland was restored to its enclosure and the helpful tourist kept watch over it until we were well out of sight to prevent any further mayhem.

Things quietened down from then on and we rode up the northwest side of the peninsula, following deserted lanes that wound through rugged landscapes. The weather was pleasant, warm and sunny, but a strong wind blew in off the Atlantic that whipped at us wildly whenever the road climbed away from the sheltered lower ground. Bare hills rose up to our right, and fell away to the sea on our left as we skirted Dunmanus bay, where small white-capped waves rolled gently over the fathomless blue depths. In the distance away to the west, the storm-weathered mountains of the Sheep's Head and Beara peninsulas stretched away into the calm expanse of the Atlantic Ocean under turquoise skies. The scene was breath-taking and I drank it in in silence, letting Dakota stride along in front and leaving Vlad, Oisín, and Spirit to their own devices far behind us.

At the end of a long day, Vlad knocked at the door of a farm and asked for a place to camp. By this time he had mastered the subtle art of securing us a stop for the night, and it wasn't long before we were unloading the horses and making camp in the corner of a field near Dunbeacon stone circle. While Vlad prepared dinner, I wandered up to visit the site.

It was quiet there among the ancient stones, many of which lay fallen on the boggy ground. There was a stillness about the land, a timeless quality, and the only sound here was of the wind whispering through the grey rocks and gorse shrubs that were scattered across the open hillsides.

I breathed in the atmosphere and the rich silence, looking out across the bay to the mountains beyond, and watching the silver crescent of an old moon rise over the barren hills.

23

The Cork and Kerry Mountains

The thing about people who have never travelled on horseback for more than a couple of days at a time is that I wrongly assume they will understand the needs of a long distance horseback traveller, while they wrongly assume that I must enjoy living in a tent, wearing dirty clothes, and not washing for weeks on end – as they did for a few short days on their trip. For these horseback holidaymakers and part time adventurers, the novelty of such discomforts has not yet worn off. They have never reached the point in a journey when, if they were offered three magical wishes, taking pride of place at the top of the list – and ranking far above sensible requests such as great wealth and eternal youth – would be a hot shower, with clean clothes or somewhere dry to sleep coming in at a close second. Those who have only tasted life on the road for a handful of days are full of the romance of the road and have yet to really endure any of its harsh realities. They like reminiscing about the joys of sleeping in wet tents and travelling in the pouring rain and are quick to forget the sheer awfulness of the experience once they're safely back in the comfort of a warm house, feeling clean and thawed out, with their equipment all nicely dried and packed away until they repeat the exercise again the following year.

Fiddler on the Hoof

I was feeling tired, fed up, and pre-menstrual on the day that we arrived at Sandra and Tim's house. Sandra was German and Tim was English and they were what the Irish call Blow-Ins – people who have blown in from abroad, accidentally landed in a remote corner of Ireland and, against all the odds, have put down roots and thrived. The couple ran a successful YouTube channel showing life on their smallholding near Bantry where Tim designed all kinds of wonderful inventions, and Sandra ran an equine assisted therapy centre. Sandra also liked filming her short journeys on horseback around Ireland. I was looking forward to staying with them. Here was someone who would understand our needs, I felt, when Sandra had offered us a place to stay. She had travelled on horseback for days at a time so she'd appreciate the hardships of the road, and instinctively know what to offer us.

It had been almost a week since we'd last seen a shower or had our clothes washed, and I spent much of the day fantasising about those simple home comforts as we rode through the hills to Bantry, and then picked our way along the busy N71 at rush hour in the direction of Ballylickey. When we arrived at our destination, however, I was dismayed to find that no shower, no washing machine, and nowhere dry to sleep were offered up.

Throughout the evening I desperately dropped subtle hints about the joys of hot showers, clean clothes, and nice, dry places to sleep, while Sandra chatted happily about her excursions on horseback over the years – mostly in the pouring rain – and Tim showed us some of the genius inventions he had made. But my hints fell on deaf ears.

I know I shouldn't have got my hopes up in the first place. Expectations are nasty things that leave you feeling sorely disappointed when they go unmet – but I'd been so certain that Sandra would have anticipated our needs that I'd allowed myself to dream.

Now if you find yourself wondering 'Why didn't she just ask?' then you, my friend, are not a true Brit, and have no understanding of how the British mind works. Of course it would have been the most obvious, logical thing to do – our hosts were friendly and hospitable people and it probably hadn't even occurred to them that we might be in want of some basic necessities. I'm sure they wouldn't have batted an eyelid had we asked for a shower and a dry corner of a barn or greenhouse to sleep in, safe from the forecast rain, but we British are a nation of socially awkward beings with some pretty weird, unspoken rules of etiquette and propriety – one of which is that you can never ask for anything outright; especially not from a virtual stranger who is showing you unrequited kindness and offering you hospitality. It's considered terribly rude!

This might seem bizarre or downright ridiculous to anyone who is not British, and unfamiliar with our ways, and you'll probably be wondering how, with this enormous handicap, we go about getting anything we need at all. Well actually, we have evolved a rather complex system of getting what we want without ever having to directly ask for it. This system relies on both parties knowing the rules, and playing the same game. It goes something like this:

The person who wants something – like a shower, or a cup of tea, for instance – carefully drops subtle hints into the conversation at appropriate moments and then waits for the other person to pick up on them. Once the hint has been picked up, the other party should offer whatever it is you want.

'That's quite straightforward,' I hear you say. But no! It's not nearly so simple. At this juncture it would still be considered much too forward to accept the offer, so feeble protestations have to be made, offers turned down with lots of 'I couldn't possibly's – which really translate to 'Yes please!' and 'This is exactly what I want.'

Both parties know this perfectly well, so the person offering must then strongly insist, turning the proverbial tables and making it now appear dreadfully rude to refuse to offer. At this point one may humbly and begrudgingly accept. The whole thing is a lengthy and overly complicated process, which probably makes no sense to anyone outside of Britain – but it's a game that all Brits instinctively know how to play and thus it works perfectly without a glitch. In this way you get exactly what you wanted all along without the awkwardness and perceived rudeness of having to actually ask for it.

There is nothing more painfully uncomfortable to a Brit than trying to engage in this carefully structured performance only to find that the other party does not know how to play, and that the only remaining option is to ask for something directly. It just isn't done!

It was in this predicament that I now found myself, with my hints falling unnoticed at the wayside of conversation and no hot shower or clean clothes in sight. Unable to break the habit of a lifetime, I simply couldn't bring myself to ask outright; and Vlad – after more than a decade in England – had become too Anglicised to be able to ask either! Anyway, he was more interested in discussing Tim's projects and inventions. In the end I gave up trying and tumbled un-showered into our tent in travel-soiled clothes just as the rain began in earnest.

It was apparent that the rain was here to stay as we packed down the sodden tent and set off in the steady, grey drizzle the following morning. I'd looked at the map with Sandra over coffee and she'd pointed out a quiet route through the mountains from Kealkill over to Kilgarvan.

My original plan had been to ride down the Beara Peninsula, which had promised plenty of mountains and some stunning views – but the clouds were low, the rain heavy, visibility poor, and the only roads onto the peninsula were busy tourist routes. It seemed like a bit of a pointless and dangerous exercise without the beautiful scenery to reward us.

I was disappointed to miss out the peninsula because Beara is steeped in ancient legends and plenty of good folk tales – one of the most famous of which is An Chailleach Bhéara (the Hag of Beara).

In older tales, she is a goddess who lived seven times and much of Ireland's landscape was created by the pebbles that fell from her apron as she roamed the country. She was the wise old crone who brought the winter and governed the dark months of the year, while her counterpart, Brigit, brought the spring and ruled the summer months. There is a rock on the Beara peninsula facing the sea, which bears the Hag's name. Some say she sits there until eternity waiting for her lover, Manannán, god of the sea, to return; but in later tales she was portrayed as a wicked old hag who tormented a priest and he turned her into stone on the spot.

Veering away from the Beara Peninsula, we headed inland to Kealkill. Beyond the imposing ruins of Carriganass castle we picked up a little lane, which wound through dark forests over the gloomy hills and eventually joined a long, gently climbing road that followed the side of a valley for many miles. The surrounding Shehy Mountains were hidden behind thick curtains of mist that rolled down bare, craggy hillsides where sparse native trees clung to rocks far from the eager mouths of foraging sheep. Little streams and rivulets trickled down the rock faces and became fast-flowing gullies that ran beside the road, before meeting tumbling rivers which flowed to the now distant sea. We crested a high pass, crossing out of County Cork and into Kerry. Here the road began a slow, winding descent, zig-zagging beneath dramatic crags etched with little waterfalls. The valley on this side of the pass was more wooded than the other, and a heavy silence smothered the grey-cloaked mountains.

We trudged happily along – in spite of the miserable, wet weather – drinking in the misty views and our wild surroundings. We were soaked to the bone but loving life and the adventure.

At last we fetched up in Kilgarvan where we found a field to call home for the night behind one of the many pubs in the village. The field was large and its sole occupants were a cow with an inquisitive young calf at foot. While Vlad made camp, I went round the field closing up the many large openings in the fence that led out onto the main road – marvelling that the cow and calf stayed put. Any animal with an ounce of curiosity and the spirit of adventure would have been long gone in the quest for pastures new; but this cow and her calf seemed quite content with the status quo, and so secure fencing clearly wasn't needed.

I had come to realise that fences in Ireland were a rather haphazard affair, with little attention given to their functionality or upkeep. The horses, donkeys, sheep, or cows that were enclosed within the insecure confines of an Irish field appeared to have reached some sort of unspoken agreement with their owners whereby they miraculously stayed put and rarely strayed, because in spite of the terrible state of the fences, we rarely encountered any wandering livestock.

Vlad and I made the rounds of the pubs that night, getting to know many of the friendly locals. By that time we were both in dire need of a wash, so much so that Vlad wondered how anybody could stand to be within a few feet of us. In the end we resorted to washing in the sink of the pub toilets, but that did little for our travel-soiled clothes.

Some of the people we spoke to suggested a route for the morning heading to Moll's Gap and the Gap of Dunloe. Those were beautiful places, they assured us, and for the most part the roads weren't too bad. In this instance we decided to listen to their suggestions because we had no real plans, so what was there to lose?

It was still raining when we left Kilgarvan in the morning. We followed a quiet road along the river, passing scattered farms, new-build houses, and fields of emerald green grass where lone hawthorn or ash trees grew.

These were fairy trees, said to be beloved of the fairy race and to act as portals to the other world. To this day, even in modern Ireland, it is considered unlucky to cut them down. Great misfortune will come to anyone who harms or destroys such a tree, or who damages a fairy rath – a circular enclosure surrounded by a wall of stones or trees. This belief is so deep-rooted in the Irish psyche that a whole motorway in County Clare was re-routed in the late 90's to avoid cutting down a fairy tree. People were worried that it would bring a curse on the motorway and incur a great number of fatalities on the road. The issue was taken so seriously that the building of that motorway was delayed for a whole ten years while it was debated.

More recently in 2017, Danny Healy-Rae – a local Kerry politician – blamed the inexplicable and continual subsidence of a road near Killarney on the destruction of local fairy forts.

The fairies of Ireland go by many names. To some they are the 'wee-folk' – small, mischievous beings who can bring both good luck or great misery on mortals who interfere with their doings. To others they are the Gentry, or the Sídhe, who some say are the Tuatha Dé Dannan – the old gods banished to another world beneath the hollow hills. In later, more Christian stories, they were fallen angels cast out of heaven. Whoever – or whatever – they are, their presence is still palpable throughout the whole of Ireland, in the lone trees, the ring forts, and the sacred hills.

24

The Black Valley and the Gap of Dunloe

It was getting late as we rode through the Black Valley. The usual continuation of houses had vanished as we'd approached Killarney National Park, following rough tracks through a deserted and mountainous wilderness of thick, boggy scrubland. The morning's rain had dispersed and the sun had emerged to illuminate rugged mountain peaks on every side – their bare slopes mottled by ever-shifting cloud shadows. We'd spent the day drinking in the wild beauty of it all.

Now we were looking for somewhere to stop, but we found no welcoming farmers to offer us a haven for the night.

'The horse people live at the other end of the valley,' a rough old sheep farmer had told us when we'd stopped to enquire. 'It's only a mile or so further along the road.'

Several miles passed and no sign of any 'horse people' could be seen. The landscape, which earlier had been so gloriously wild and uninhabited, now appeared bleak and unwelcoming as we trotted hard along the road, searching among the rocky terrain for a field or somewhere suitable to camp – but we found nothing.

In a pull-in by the river, we came upon a tubby little man with long grey hair and a long grey beard who, had he not been sporting a brightly coloured tie-dye t-shirt, would have strongly resembled a Tolkienian dwarf. He was making a fire next to his trailer, on which was painted the image of a wild boar. He looked like he could tell us a good tale or two – but we needed to find a field for the horses, so we pushed on.

At long last we found a nice man named Gene who said we could camp in his field beside a fast-flowing river. To my delight, he also offered us the use of a shower in his empty holiday cottage a little way down the road. I leapt at the kind invitation. Vlad, on the other hand, opted to wash in the freezing cold river next to the camp while he cooked our dinner on an open fire. He was in his element!

A few miles away, across the mountains from where we were camped, lay Lough Leane. It is said to be an entrance to Tir Na nÓg – the land of youth, and the Celtic otherworld inhabited by the Tuatha Dé Dannan. One of the best-known stories of Tir Na nÓg tells how Niamh, daughter of Manannán, appeared to the Fianna one day while they were hunting on the shores of Lough Leane. She seduced Oisín, son of the legendary warrior Fionn MacCumhail, and lured him away to live with her in the land of the ever-young.

After three years in the otherworld Oisín grew homesick for Ireland; he longed to see his father and all his friends again. Although Niamh implored him not to go, Oisín was adamant. In the end, powerless to stop him, Niamh gave Oisín her father's magic horse, warning him that whatever he did, he must not set foot on Irish soil.

Oisín rode across the waves and at last he found himself in Ireland – but it was not the country he remembered. Three hundred years had passed there since Oisín had left for Tir Na nÓg. Gone were the Fianna and all the places that were familiar to him. Instead he found a Christian land full of churches and monasteries.

As Oisín roamed the countryside, he came across some men trying to move a heavy stone. They called out to him for help, but as he reached down to roll the stone, the girth broke on his saddle and he fell to the ground. As soon as he touched the soil, his youth faded and he withered into an old man. The horse disappeared, and Oisín never saw Niamh again.

Oisín lived out his final years among the Christian monks where, it is said, he met Saint Patrick himself, and told him tales of the Fianna and of ancient Ireland.

Many old poems of Irish mythology are attributed to Oisín, and he is regarded as one of Ireland's greatest bards.

We were awoken the next morning by the sound of clattering hooves. Looking out of the tent, we saw ponies pulling traps flying along the road on the far side of the river. We packed up our gear and rode through a steady stream of tourists up to the dramatic Gap of Dunloe.

The Gap of Dunloe is a narrow pass which runs between two mountain ranges: the MacGillycuddy's Reeks and the Purple Mountains Group. The scenery there was spectacular, with tall mountains rising on either side of a snaking road. Little stone bridges crossed and re-crossed a tumbling stream that, halfway down the valley, fed into a series of lakes whose dark waters looked as though they contained all kinds of unimaginable monsters – but on that day we had more pressing things to attend to: Oisín needed new shoes.

His last set had been on since just after we'd left Gloucester and, almost four hundred miles later, they were looking rather the worse for wear. We flagged down a passing pony and trap and asked the driver whether he could help us find a farrier.

'Ask the lads down at Kate Kearney's Cottage,' he said, cheerily. 'They'll sort ye's out!' And off he sped.

Kate Kearney's Cottage was a bar at the foot of the pass and outside it, in a large, open yard, stood a great number of cobs tied to hitching posts next to piles of harness and small wooden traps called jaunting cars. The trap drivers, known as jarveys, milled about, talking amongst themselves and smoking cigarettes while they waited for the next wave of eager tourists to arrive.

Oisín was closely inspected by the jarveys and was met with great approval. They thought him a fine specimen of a horse and asked us many questions about him: what was his breed, was he broken to harness, how much was he worth, did he have good stamina, and the like. He was just the sort of horse that these men liked. Dakota, on the other hand, was largely ignored. He was deemed too finely built, and above all, he was skewbald. Only the Travellers have coloured horses in Ireland, they said.

One jarvey called Bernie rang a farrier who said he would come out to us that night, if we told him where we were staying; but we didn't yet know. The kind jarvey then gave us the number of a man named Eric who said we could camp in his stables near Castlemaine at the head of the Dingle Peninsula. We rang the farrier back, and he agreed to meet us there that evening.

Castlemaine was still a good thirteen miles away so we didn't hang about and set off down the road towards Beaufort, past the Dunloe Ogham Stones. Here the land levelled out to flat, green pastures full of beef and dairy cows and small flocks of sheep. The regular smattering of unseemly new-build houses in every field resumed, and the rugged mountains were left far behind us.

Our host, Eric, met us in Castlemaine and he told us there were another five miles to go to his yard at the foot of the Slieve Mish Mountains. Eric offered to take Spirit and all our gear on ahead so that we could ride unencumbered for the last stretch. It was growing late now, so with barely a hesitation, we threw our faithful wolf and all our possessions into the back of a stranger's van, and watched him vanish down the road.

At a normal pace the five miles would have taken us just under an hour and a half, but daylight was limited and Oisín still needed to be shod, so we trotted the rest of the way and arrived on the yard in record time.

Although she'd barely been parted from us for an hour, Spirit acted as though it had been many months. She was both delighted to see us, and resentful that we'd been so quick to abandon her. She whined and yelped, leapt about us wagging her tail, and pawed at us with her long, sharp claws.

The farrier arrived shortly after, wrestled with the strongly objecting Oisín to get some new shoes on, and then we bedded down for the night on the floor of a stable and fell into a deep, contented sleep.

25

Dingle

Cú Chullain, son of Lugh, sat beside a stream sharpening his sword and watched as the first grey light of dawn broke on the horizon. Men stirred in the camp around him, readying themselves for a fight, and in the distance a raven croaked hoarsely. Blood would be shed before the sun rose.

Suddenly the water beside him began flow a milky white. It was the signal he'd been waiting for.

The men crept quietly up the mountain through the trailing mist. Ahead of them loomed a dark fortress built all in stone. The door was unguarded and stood ajar. This would be all too easy, Cú Chullain thought, as he pushed the door open and entered the castle, sword held at the ready.

In the shadows a figure moved. He could just make out the shape of a woman – Bláthnat. She beckoned him quietly and led the men to where the sorcerer, Cú Roí, lay sleeping. It was a deep, peaceful sleep – as innocent as that of a child.

Cú Chullain pounced and with one swift movement drove his sword home to his enemy's beating heart.

The day was blustery, with intermittent showers rolling in off the Atlantic as we set off down the Dingle Peninsula heading for Inch. The Slieve Mish Mountains, which rose up to our right, were hidden behind a blanket of low hanging cloud.

Somewhere up in those gloomy heights stood the ruins of Cathair Chonroi – The Fort of Cú Roí – an old castle where centuries before an ancient tale of betrayal and revenge unfurled.

The fortress had belonged to a great sorcerer named Cú Roí, who was also a warrior and king of Munster. In a tale from one of the Irish mythological cycles, Cú Roí joined the great warrior Cú Chullain on a raid of Inis Fer Falga – modern day Isle of Man. They stole cattle, a magic cauldron, and the king's daughter, Bláthnat.

Whilst dividing the plunder, the two warriors fell to arguing when Cú Roí demanded Bláthnat as his share of the spoils. Cú Chullain refused because he and Bláthnat had become lovers. Enraged, Cú Roí seized Cú Chullain and stuck him in the ground, burying him up to his armpits. He cut off Cú Chullain's hair, then took Bláthnat away to his home at Cathair Chonroi.

Seeking revenge for this humiliation, Cú Chullain travelled to Cathair Chonroi to kill his enemy and retrieve his lover. The fortress was heavily protected by magic, and guarded by many men, so Bláthnat had to help Cú Chullain in his quest.

She sweet-talked the sorcerer Cú Roí, flattering him, and saying that a man as powerful as he should really have a bigger castle. Cú Roí agreed and sent all his men away to fetch stones for the building of a larger fortress. That night, while her husband lay sleeping, Bláthnat gathered up Cú Roí's weapons and hid them safely away. Then she poured milk into the river, sending a signal to Cú Chullain and his men to say that it was safe to attack.

Cú Chullain, seeing the signal, stormed the castle and killed Cú Roí.

In another version of the tale, the sorcerer Cú Roí had hidden his soul in an apple inside the belly of a salmon that lived in the river below the fortress and only surfaced once every seven years. In that version of events, Bláthnat discovers this, and tells Cú Chullain who finds the salmon and slaughters it, thus killing his enemy.

Neither version of the tale ends well, however, as Ferchertne – Cú Roí's poet – is so angry at Bláthnat's betrayal that he grabs hold of her and leaps off a cliff, taking them both to their death.

Dingle

When we arrived at Inch, Mamood — the owner of Sammy's bar — let us pitch up in the campsite beside the beach and turn the horses into the paddock above.

Inch beach — contrary to what its name suggests — is a sandy spit that stretches for three miles into Dingle Bay. No self-respecting horse rider can turn down a three mile long sandy beach, so once we'd set up camp we fetched the horses, tacked up once more and headed down to the strand.

We charged along the beach, deafened by the sound of crashing waves, and battered by the wind and the spray. The horses flew, deftly dodging the large piles of green and yellow sea-weed that had washed up at the last high tide. Both were alert, on edge, and game for a gallop; they raced each other along the beach.

Across the wind-whipped, white-capped waves, we caught glimpses of the Iveragh peninsula through curtains of rain that moved over the bay. Dark clouds obscured the tall peaks there, and mist tumbled down steep mountainsides then lifted with the turbulent gusts of wind. Storm clouds scudded before the gale, breaking every now and then to reveal patches of bright blue sky — serene above the turmoil of the wild waves. Long fingers of golden sunlight spilled through these gaps in the shifting clouds, setting the spray and white mist a-glimmer and illuminating the turbulent dark water, all crowned with white foam. There was a stark contrast here between light and dark, stillness and motion. The whole scene was elemental and wild.

We tried to get the horses into the sea, but they shot backwards before the encroaching waves that crashed onto the beach — snorting their displeasure at the swirling water which lapped at their toes. Of the two, Dakota was the more brave and finally he dared to enter the water and wade among the breaking waves. At last, red-faced, and wind-battered we headed back to camp, put the horses in their field, and made for the crowded bar to sample the Guinness and chat to the many tourists there.

The next day we continued on down the peninsula, picking up lanes that wound through the boggy hills and avoiding the main road wherever possible. Low clouds still obscured the surrounding mountains, but what little we could discern promised beautiful countryside. It was early evening when we reached our destination – Camphill Community Dingle, where some friends of my mother had offered us a place to stay.

Camphill Communities were a place adults with special needs lived alongside people without disabilities, sharing a home and daily life as one large family. I had been brought up in a Camphill Community, living in a house with five adults with special needs. I had lived there from the age of seven right through until I left home at nineteen. There's something unique about those communities; a certain atmosphere that makes them feel very comfortable and homely, unlike the sterile, soulless care homes that you normally find. Sadly, however, in recent years red tape and bureaucracy had caused the demise of these communities, stripping away their ethos and forcing them to become more 'mainstream'. Instead of living in a house and sharing every aspect of their lives with the residents, 'staff' were now forced to live off-site and come in to work shifts. Where once the community had supported the needs of all its inhabitants without salaries or hourly pay-rates to quantify the work, everyone now had set working hours and fixed wages. The sense of family and vocation was gone, and as a result the communities were being eroded, slowly and surely, to become just run-of-the-mill care homes.

Upon arrival we were greeted by some of the residents, their support workers, and several old friends of my parents – the last stalwart survivors, clinging on by a thread, and trying desperately to keep the spirit of Camphill alive before their way of life was stamped out altogether. They had founded this community many years earlier and had worked hard to build it from the ground up, dedicating the best years of their lives to it. It was hard for them to see everything they had worked for being destroyed by strict regulations, imposed on them by an overbearing system which didn't understand Camphill and therefore couldn't appreciate its unique and holistic approach to working with people with special needs. Sadly the ones most affected by these changes were the residents themselves, whose quality of life was now being impoverished.

Once the horses were settled in their field, we joined everyone for a meal in the large building that served as library, weavery, and

Dingle

needlecraft workshop. The community was celebrating the birthday of one of the young German girls who had come to spend the summer volunteering there.

After dinner, some of the support workers and volunteers invited Vlad and me to go with them into Dingle town. There would be live music, they promised, so we leapt at the chance.

In Ireland dogs are generally not allowed in pubs or bars. In fact they are rarely even allowed in houses. Mostly, they live outside in kennels, spending their days running loose and doing whatever they like. It's a much freer existence than dogs have in Britain where they spend most of their day locked in a house and only get a few meagre walks a day. Irish dogs are sociable creatures and you'd frequently see one trotting down the road to the next house along to pay a visit to the neighbouring dogs. Sometimes you'd see a small pack making the rounds, or heading off to play together – and never a supervising human in sight.

Because we doubted whether we'd be able take her into any pubs with us, we made Spirit a nice bed in a potting shed, tied her long lead to the wall, gave her a bowl of food and water, and firmly bolted the door. She seemed content enough as we left.

In Dingle we explored the little streets and found live music sessions happening in several of the pubs. Eventually we settled on one where two men were sat in a corner playing traditional tunes on the fiddle and uilleann pipes. Polkas seemed to be the music of choice here and we gathered a couple of local tunes.

At midnight the music stopped, the tourists left, and the bar was turned into a nightclub. It quickly filled with young, drunk people dancing to thumping modern music that grated on my nerves and brought back hideous memories of when I was eighteen and used to enjoy that sort of thing.

'I want to go!' I had to shout for Vlad to hear me over the music. 'This is awful!'

'Let's stay for a bit,' Vlad pushed. The young volunteers seemed to be enjoying themselves.

'I want to check on Spirit,' I insisted, knowing that by now she'd probably have had enough of the potting shed.

It wasn't until about an hour later that I could take no more and insisted we leave. Vlad and I shared a taxi back to the community with another young German volunteer named Rebecca, who was also not much of a party-goer and didn't like the music either. As the taxi pulled up outside the community we were greeted by a little grey wolf who materialised in the road, looking very pleased with herself indeed. With sinking hearts we went to see what havoc she'd wrought on the potting shed.

In the grand scheme of things, I suppose a chewed up lead, a mangled electric cable (which luckily hadn't been live), a few gnawed table legs, some ripped bags of fertiliser, a pile of strewn potatoes, a badly scratched door, and one broken lock were small damage and all easily fixed. Luckily Serge, the lovely French gardener, seemed to find the whole thing quite amusing in the morning when we confessed to the damage.

We spent a few days resting in the community, feasting on fresh vegetables from the extensive biodynamic gardens, getting to know the staff, residents, and volunteers, and joining in with the community's daily activities. After two days' rest, however, it was apparent that the horses needed to hit the road. We knew this from the way that Dakota was performing acrobatics in the field, with Oisín charging about beside him, bucking and farting loudly. Clearly the rich, Irish grass was going to their heads and they needed some work to let off steam!

By the time we'd reached Dingle, my walking boots had practically disintegrated and were well beyond repair. Luckily the Camphill Community had a large store of wellies and Serge told me to help myself. I found a good pair that fitted well and adopted them for the rest of the journey.

When we left the Camphill Community, we made our way through Dingle town, heading for the Connor Pass. It was one of the highest passes in Ireland and the views were spectacular, everyone assured us - except that when we got there everything was shrouded in thick, wet mist and we could barely see two yards ahead of us on the narrow road that wound through the boggy hills. As we began our descent, I was aware of a sheer drop on our left beyond a very low stone wall

Dingle

and I flinched every time a car came hurtling round the bends and narrowly squeezed past the horses in the blinding fog. It was nice to reach our destination unscathed and in one piece.

Serge, the gardener from the Camphill Community, had invited us to stay at his home not far from Castlegregory. Once the horses were happily turned away in a nearby field, we sat down to a delicious dinner of home-grown vegetables and freshly baked sourdough bread. That night we slept in the converted out-building where there was a comfortable bed and an en-suite bathroom with a shower! It was the first bed we'd seen since leaving Wales almost three weeks earlier, and was a much-welcomed luxury.

In the morning after breakfast we loaded the horses up again and Serge pointed us towards the beach. If the tide was out we could ride the whole way along it and never hit a road for nine miles from Castlegregory to Derrymore, he said.

To our delight, the tide was out, the clouds finally lifted, the sun shone, and a lovely relaxed day was spent cantering across the white sand, splashing through shallow rivers as they fed into the sea, and admiring the spectacular scenery of the Dingle peninsula. It was sheer bliss!

26

Travellers in Kerry

'Oi! Stop! Where do you think you're going?' The security guard was red-faced and angry.

'To Halfords,' I replied, slightly taken aback at this flood of unwarranted antipathy. 'I need some gas for my stove.'

We had been making our peaceful way through the car park of a large retail park in the centre of Tralee when an irate security guard had come marching up behind to halt our progress.

'You can't do that!' he spluttered, angrily.

'Why not?' I asked. 'We'll be two minutes. We literally just need some gas.'

'What'll ye do with the horses?'

'Tie them up here,' I said, coming to a halt outside Halfords next to some conveniently placed railings.

'You can't!' he repeated, aghast. 'What about the cars?'

What about the cars? I wondered, looking around at the half-empty parking lot.

'I won't be a minute,' I said, tying Dakota and Spirit to said railings and determinedly ignoring the guard's fervent protestations. 'Keep an eye on the horses, and tell this gentleman what we're doing,' I said, turning to Vlad, and left him to deal with the angry security guard while I went to find gas. He's much more of a people person than I am, so I felt confident he'd have the whole thing sorted out by the time I returned.

Sure enough, when I re-emerged from Halfords, gas in hand, Vlad had worked his magic. The security guard had calmed down, and the two were having a nice, friendly chat about our journey. The guard apologised for being so unreasonable earlier, explaining that his boss had been with him and they'd thought we were Travellers, so he'd had to look like he was 'dealing with the situation'.

The guard showed us a shortcut out of the shopping complex, wished us luck with the journey, and then wandered off to worry about Traveller invasions elsewhere.

We'd heard a lot about Travellers in Ireland, and been mistaken for them, too. Mostly we'd encountered a lot of deep-rooted prejudice against them. They were gypsies and tinkers – a lawless people who abided by their own rules and customs, kept themselves to themselves, and lived in tight-knit communities where even the police dared not set foot. Be careful, people warned, they were thieves and they'd rob you blind.

It felt like a rather sweeping and negative generalisation of an entire group of people, most of whom, I felt sure, were as decent and law-abiding as the next person.

The exact origins of the Irish Travellers are unknown and much disputed. Some say they are the descendants of the ancient kings of Ireland, or of an old, pre-Celtic race of pastoral nomads; others say their ancestors were a class of travelling tradesmen; and according to other theories they were people forced to a life of nomadism by war or famine. Because the Travellers lack any written history of their own, these theories are hard to confirm or refute and perhaps there is a grain of truth in all of them. All that is known is that for centuries, groups of nomadic people wandered the land living in tents or horse-drawn waggons, carrying news from county to county, and eking out a living as best they could. Some were skilled craftsmen, or tradesmen; others travelled with the fairs and circuses; and many were good horsemen and dealers, reputed to have a way with animals. It was widely believed that the Travellers and the Gypsies were possessed of supernatural powers and could read fortunes or put curses on those who crossed them.

In the 1960's, the Irish government took steps to deal with the large itinerant population in the country and introduced laws criminalising nomadism and trespass. Traditional halting places were eventually closed off by local councils and the Travellers were forced to become settled, or were restricted to a handful of designated sites between which they could travel – now in vehicle-drawn caravans.

Recent studies on the Irish Traveller population show that, although they are of Irish decent and unrelated to the Romani Gypsies – with whom they are so frequently confused – they are genetically distinct from the rest of the Irish population. Thanks to this study, in 2017 Irish Travellers were finally recognised as an indigenous ethnic minority with their own customs, taboos, languages, and folklore. We'd been told that the Travellers still had the best storytellers in the land and that within their tight-knit communities they kept the tales of old Ireland alive.

We were just about to make good our escape from the tarmac desert of Manor West Retail Park when a soft-spoken man with long, greasy brown hair, staring blue eyes, and an earnest expression came over to talk. He took great interest in the journey and the logistics of living on the road, and said he travelled a lot, too. A pleasant chat was had, but as he turned to leave, he suddenly said:

'I hope you don't have to wait till you're in trouble to call out to the Lord Jesus. He can save your souls from an eternity in hell, you know.'

'Oh ... right! Erm, thank you!' I said, rather taken aback because I didn't really know how to respond to that one.

We were barely a quarter of a mile down the road when the man caught us up again as we were trying to navigate a roundabout and gave us a generous donation for our charities.

'What's your name?' I asked, making a note of the donation on my phone.

'The only name you need to know is Jesus,' he said, fixing me with his staring blue eyes before he hopped into his car and vanished, so I made a note that Jesus had given us €20 for our charities that day.

We had quickly discovered that the Irish were an incredibly open, friendly people, who loved nothing more than to stop and chat to find out what we were doing. When they heard that we were raising money for charity, most people we talked to would immediately dig into their pockets to give us a generous donation, and thus most of the money we raised on our ride was from Ireland.

To get out of Tralee we had to follow a very busy N-road up to a roundabout, using the wide hard shoulder to avoid the speeding traffic. At a roundabout we picked up a cycle track which ran next to a dual carriageway, before we finally struck out along a quiet road that led through acres of forestry plantation and peatbogs all the way to Knocknagoshel where my old friend Louis lived.

We spent a night at his house and in the morning we continued on through rural Kerry. Here almost every house we passed was decorated with green and yellow bunting – the Kerry colours. All were showing support for the local Gaelic Football and Hurling teams. Huge green flags flapped in every gateway which at first caused poor Dakota some consternation, but after the first twenty or so they lost their terror and became an accepted part of life.

We'd been warned that finding places to stop might become difficult as we neared Limerick. It was a rough place, full of Travellers, and we were close to Rathkeale, too, which is renowned for its Traveller community. People would be suspicious of us, especially with a coloured horse in tow, we were told. But we encountered no such problems.

Near Monagea a man named Kevin invited us to camp in his garden when we asked if he knew of anywhere nearby that we could stop for the night.

'There's a field out the back. Work away, lads, work away!' he said cheerily. 'You're not going to murder us in the night are you?' he suddenly added as an afterthought.

We rigged up a make-shift paddock with electric fencing in the field out the back, pitched our tent in the garden, and then joined our hosts for some dinner.

Kevin lived in a large, new-build house with his Latvian wife, Kristiana, and their young children who adored the animals and fed

Spirit about five meals' worth of dog food. She thought she was in heaven, until she vomited it back up several hours later.

The following day we trekked along busy roads through flat countryside, skirting the foot of Knockfierna – the Hill of Truth – beneath which the king of the Munster fairies is said to reside.

A local folktale tells how a hunchbacked fiddler was walking home from a dance one night when, passing the hill, he spied a group of little people dancing in the moonlight. They had no music for their revelry, so he sat down on a rock and began to play. The fairy crowd was so delighted with his music that, as a reward, they removed his hump.

Word got out about what had happened, so another hunchback – also wanting to rid himself of his hump – set out for the hill on a bright, moonlit night. Sure enough, he saw a crowd of small folk dancing without music.

Sitting down on a rock, he took out a fiddle and began to play. This man, however, was not such a skilled fiddler, and the noise he made was so awful that the fairies gave him the other man's hump in punishment for his terrible playing.

As tempting as it was to stop and try my luck with my own fiddle, I decided it was just too risky.

Near Crecora a pleasant couple invited us to camp in the field next to their house and in the morning we headed for Limerick, straight into the town centre. Why? I hear you ask. Because sometimes it's easier to just go right through the middle of a town. Although they can be busy places full of people and traffic, there are plenty of handy pavements, nice small side-roads, and speed restrictions, too. It certainly makes the going simpler and safer than trying to navigate the bigger, faster roads around the outskirts.

Vlad rode and I walked, and we got lots of funny looks as we picked our way through the town – manoeuvring roundabouts, bus-lanes, cycle paths, traffic lights, and a few pleasant side streets. Finally we crossed the river Shannon into County Clare and headed out of the city on a long, straight road towards Broadford. Suddenly the heavens opened and it poured with rain so we pulled over to don our waterproofs. By the time we'd done this, the rain had stopped, the sun shone again, and our waterproofs made us hot and sweaty – but when we went to take them off, the next heavy rain shower was upon us. We were faced with the uncomfortable dilemma of being dry from the outside but hot and soaked with sweat from the inside, or being sweat-free but cold and soaked through from the outside. Neither was a particularly pleasant option.

Several of the local Travellers driving past us on that long stretch of road stopped to find out what we were doing. They asked if we needed anything, and offered us assistance and a place to stop and rest. Here were the dreaded Travellers about whom we'd heard so many awful things – and ironically, they were among the friendliest people we'd met in Ireland so far, going out of their way to offer us help and hospitality.

Nearing Broadford the landscape became wilder and more undulating and we found ourselves wandering along little lanes over near-deserted hills. Large areas of forestry were dotted about these mountains and the non-native Sitka spruce and Douglas fir were planted so thickly that no light ever reached the forest floor. Nothing grew beneath the thick canopies but moss and fungi, and the hills were eerily silent because even the wild birds could find no sustenance in that barren forest desert.

As we crested the brow of a gentle hill, the road ran alongside a tumbledown stone wall. A herd of scruffy ponies and donkeys in the enormous field the other side came over to get a closer look at the strange spectacle we presented. In amongst them I spied some creatures who were neither ponies nor donkeys. They were short and stocky, about 13hh tall, their coats were a lovely dappled grey in colour, and they had large heads with beautiful long ears. They were mules! A cross between a horse and a donkey, mules are reputed to be hardier, more intelligent, and a whole lot more difficult to train than horses. It is said that all mule-handlers can work with horses, but not all horse-people can work with mules. Could I work with a mule? I wondered, idly, admiring the beautiful creatures before us. I'd always fancied having one. There's something endearing about those long ears and solemn faces, and if you can gain their trust they become a faithful friend for life.

I said as much to Vlad. It was only a passing comment, a fanciful idea for some time in the distant future.

'I'll get you a mule!' Vlad cried – ever enthusiastic and woefully lacking in well-grounded realism. 'Anything for my queen!'

I rolled my eyes. We were halfway around Ireland. My crazy young horse had only just settled down and become sane and sensible. I certainly didn't want to shake things up by acquiring another creature now – particularly not one which had a reputation for being difficult. That would be ridiculous.

I was quite relieved when, stopping at the house next to the field to ask about a place to camp, we found no one home and so had to push on.

Phew! Crisis averted, I thought, and a short way down the road we found a field to stop in for the night. I set up camp and made dinner while Vlad went off to chat to our hosts, Joe and Liz. He returned several hours later as the light was fading, looking very pleased with himself indeed.

'I've got you a mule!' he said with delight. 'We'll go and see it in the morning.'

27

Micheál

'One hundred euros and he's yours!' Micheál (Mee-hawl) exclaimed.

We were standing in a shabby, run-down yard in the pouring rain, and in front of us was a filthy pen containing a large donkey and a tiny little black creature who was barely discernible against the shadows. Here was the mule that Vlad had arranged for me to see on the previous evening. Micheál, who was the owner of the ponies, donkeys, and mules we'd seen on the mountain, also happened to be the father of our host, Joe, from the night before. We'd come up to the yard after breakfast and introduced ourselves to Micheál – a scruffy, thickset man who looked to be well into his sixties, and seemed rather hard of hearing.

'So, it's a jennet you're after, is it?' he said loudly, eyeing us up. 'I've got one ye can have.'

A jennet, we discovered, is more commonly known as a hinny: the offspring of a male horse and a female donkey (a mule being the offspring of a male donkey and a female horse). In my ignorance of these matters I made no such distinctions, and simply called them all mules.

Actually, we were not after anything at all. It was an absurd idea to take on another animal, but Vlad had made all the arrangements, so we'd had to go and see what was on offer. All that remained for us to do now was look at the creature, say no, sorry we're not interested, and that would be that. It was simple.

Micheál led us into the yard and pointed at a filthy pen in a ramshackle old building.

'One hundred euros and he's yours!' he'd said, after first telling us that a mule had made twelve hundred euros at a sale earlier in the year, and another had gone for eight hundred. I had the impression that we were supposed to consider this jennet a bargain at a mere one hundred euros.

I peered into the darkness of the pen, trying to get a better look at the creature that stood hidden behind the enormous donkey. It was black on black against the shadows, but I could just make out two long white socks on the hind legs and two long ears. This was definitely not one of the beautiful dappled mules I'd admired in the field the day before. At barely 10hh tall, this creature was tiny – much too small to be of any use to us anyway.

'What's wrong with his eye?' I asked, squinting into the shadows to try and get a closer look. I could only make out one eye, carefully watching us. The other was missing! ...No, not missing, I realised, as I adjusted to the darkness; his right eye was collapsed, crumpled into its socket.

'Ah sure, that doesn't affect him at all!' Micheál assured us, confidently.

'What about those mules up in the field?' I wanted to know. 'Are they for sale?'

If I was even going to consider taking on another equine then I'd much rather one of those. At least they were big enough to be useful. And they had two eyes.

'You wouldn't be wanting one of them, now.' Micheál shook his head gravely. They were unhandled, entire, and still too young, he added. 'This one will do ye grand.'

'How old is he? Is he handled? Castrated?' I wanted to know – still strong in my conviction that he really was too small to be of any use to us. We just had to tell this rough old horse-dealer we weren't interested and walk away. Why was it proving so difficult?

Micheál

Micheál said the jennet was five years old, he was castrated but had no passport, he had a headcollar on, and he'd acquired him in a job-lot deal with a local farmer. Beyond that, he knew nothing.

'Look, do ye's want him or not?' Micheál was getting impatient.

'Can you catch him? I want to see him in the light,' I insisted, my curiosity starting to get the better of me now.

Micheál looked like this was asking a bit much, but eventually he grabbed a length of rope that hung on a nail in the wall, opened the pen gate, and let the donkey and mule into the yard. A scuffle ensued as he chased them both into a building and down a passage where he managed to corner the terrified mule and – deftly avoiding the wildly kicking hind legs – strung a rope through its head collar and hauled the trembling creature back out into the yard.

Clearly the poor animal hadn't been handled much – and certainly with no kindness or gentleness! Nope, I decided firmly, I definitely didn't want to take on a semi-feral, ridiculously small, one-eyed mule to cart around Ireland with us. That would be stupid.

I looked to Vlad to confirm my resolution and be the sensible voice of reason here for a change, but – having landed us in this awkward situation – he was now being very non-committal, and had backed off entirely, leaving any final decisions up to me.

I looked to the jennet next, who had finally stopped trembling and was now studying me closely with his one beady little eye. My heart melted a little. Poor mite. What sort of future would a tiny, half-blind creature like this have? What hope for a good life? My strong resolve to leave this animal here and walk out of the yard without so much as a backward glance or a second thought to his fate was weakening fast.

'I'm going out in a minute. Do ye's want him or not?' Micheál wanted a sale, and quickly. We were wasting his time with all this faffing about and indecisiveness.

'Look, we're going to be stopping for a week or two with my mother over in Mountshannon. Can I think about it and come back to you?' I said, in a last-ditch attempt to hold onto the final shreds of my good sense and reason. If I weren't standing in front of the poor mule the decision to abandon him to an unknown fate would definitely be easier.

'If ye don't have him today, I'll be taking him to market on Saturday,' Micheál said firmly.

Heart-over-head decisions are my absolute speciality when it comes to animals. Sense and reason are not. I'd acquired all my horses on a whim as impulsive decisions, void of all logic and rationale, and always against everyone's better judgement – including my own. So far that approach to acquiring equines had worked out just fine, so why shouldn't it this time? Dakota was settled and relaxed now, life on the road had become quite easy – boring even! Perhaps it would be fun to throw an unknown quantity into the mix, add another dimension to the adventure – just to keep us on our toes. Anyway, he could help carry the packs!

'What do you think, little man? Do you want to come with us?' I asked the creature who was still watching me intently. He shook his head violently from side to side, his long ears flapping in a most endearing manner.

That was not the reaction I was looking for. I tried again.

'Do you want to have an adventure all around Ireland?'

Again, the mule shook his head. This was probably a bad sign, but I laughed it off.

'I'll give you thirty euros for him.'

'Fifty and he's yours,' Micheal replied, holding out his hand to shake on the deal.

I sighed, shook the proffered hand, and begrudgingly coughed up fifty euros. Somehow I managed to decline the offer of the donkey for a mere fifty euros more. That would definitely have been pushing the boundaries of sanity.

'May he bring ye luck!' Micheál called cheerily after us as we half led, half herded the little creature out of the yard, and I suddenly wondered, too late, if I had made a terrible mistake.

'What shall we call him?' Vlad asked as we chivvied our newly-acquired companion back towards the field where we'd left the horses grazing.

At that moment, the little jennet opened his mouth and let out the most hideous yet plaintive noise I had ever heard. It was like a cross between a donkey's bray and a horse's whinny: 'Meeeeeeeee-haw-haw-haw-haw.'

'I think we'd better call him Micheál, as a reminder of where he came from,' I replied. 'It doesn't get more Irish than that ... and anyway, he can say his name!'

We had an interesting day covering the sixteen miles from Broadford to Mountshannon where my mother lived.

I'd called her as soon as we got back to the field and sheepishly confessed to our latest acquisition. For some reason my mother was never too keen on my equine purchases but, after some admonishment, she agreed to come out and pick up all our gear to save us having to carry it that day. She also took Spirit, because we weren't entirely sure how the mule would respond to his new life, nor how the horses would take to the mule, and the fewer things we had to worry about the better.

I walked with Micheál in front of the horses because he didn't walk well behind them – digging his heels in and refusing to budge – and Vlad rode Oisín and led Dakota.

Micheál very quickly impressed me with his calm and level-headed approach to life, and to being led – which I felt was something of a new concept for him. I doubted he'd had many dealings with vehicles before either, and I was apprehensive when I heard the first few cars approaching: they'd have to pass him on his blind side and I wasn't sure how he'd react. However, rather than lose his head and bolt off down the road in terror, as most horses would do, he stopped, assessed the situation in a thoughtful manner, and then reacted accordingly – which in most cases meant doing nothing at all because nothing we encountered was deemed a great threat to his immediate safety. He was the complete opposite of Dakota, and it made a refreshing change.

Micheál had a crash course in leading, dealing with cars, lorries, traffic cones, flapping tape, drain covers, and painted lines in the road, all in one day – and all were met with an admirable amount of measured good sense. Apparently flying off down the road was not one of his favoured reactions – for which I was deeply grateful, because although tiny, he was a strong little beast. He tripped merrily along beside me and we became properly acquainted throughout the day. I talked to him softly, praising him a lot, and promising him a life of adventure and world exploration – none of which seemed to inspire him quite as much as I'd hoped. He kept his good eye on me the whole way, and for the most part he seemed content. He was certainly an intelligent, discerning little creature, and a fast learner. We'd have him tamed and trained and carrying packs in no time, I felt sure.

The horses, on the other hand, did not impress me nearly so much with their behaviour. Dakota was scared of the little, long-eared jennet and kept snorting at him suspiciously with the whites of his eyes permanently on show, and Oisín – ever protective of Dakota – adopted his best war-horse stance and tried to charge in for a full-scale attack at regular intervals. I have to say, Vlad coped remarkably well with both of them in their respective reactions to the newest member of the herd.

We were halfway along the main road between Broadford and Bodyke when Micheál spied a horse in a neighbouring field. Up until this point he had been very quiet, but – being a sociable little Irishman who likes nothing better than to pass the time of day with all and sundry – he stopped dead, opened his mouth, and emitted a long, deafening, high-pitched wail.

Oisín and Dakota both took fright at the terrible sound and shot off down the road at a gallop. Somehow Vlad managed to stay put, got them both back under control, and persuaded them to turn around and come hesitantly back towards the perpetrator of that dreadful noise; Oisín with a renewed enthusiasm for killing the little creature once and for all.

Thus the day passed with the horses shooting off whenever Micheál stopped to talk to a horse or donkey, Oisín making regular attempts on Micheál's life, and Micheál having a baptism by fire to an itinerant life far away from all that was familiar to him.

We passed through Bodyke, and then Tuamgraney. From there we followed the unbroken line of houses which at some indistinct point became Scarriff. We stopped at the vets in the centre of town to enquire about getting a microchip and passport for Micheál, and we gave him a wormer for good measure. Then we plodded out on a quiet back lane, which ran between pleasantly undulating green pastures along the foot of the Slieve Aughty Mountains for the last few miles to Mountshannon. Gaps in the rolling hills afforded us views across the still waters of Lough Derg towards the distant mountains of Tipperary, and every now and then we'd catch a glimpse of Holy Island's round tower peeking out above the treetops.

Holy Island, on the western edge of Lough Derg, was an ancient monastic site where the ruins of an iconic round tower and several churches dating as far back as the 8th century A.D. can still be seen today. As with many of the other Celtic monastic sites found throughout Ireland, Holy Island had been subject to several raids between the 8th and 11th centuries when bands of marauding Vikings had wrought terror on Ireland – plundering and pillaging everywhere they went and leaving nothing but a trail of blood and burning settlements in their wake. In 1014, the Vikings' reign of terror was finally brought to an end when the legendary High King Brian Boru led armies against them in the Battle of Clontarf.

Brian Boru, who is still hailed as a national hero in Ireland, was born in Killaloe at the southern end of Lough Derg and there is an oak tree near Tuamgraney which is known as the Brian Boru Oak. At more than a thousand years old, it is one of the most ancient trees in Ireland.

At long last we arrived at our destination, tired and aching, and were re-united with the very reproachful Spirit. She had thought we'd abandoned her again and had spent the entire day crying and whining, and giving poor Mum no peace at all.

Mum had organised a field for the horses with some neighbours – Tonia and Nard – who were blow-ins from the Netherlands and Germany. They'd lost their horse some time before and were more than happy for the field by their house to be grazed down again. I hastily constructed a small paddock for Micheál to keep him safe from the still murderous Oisín, and then let him loose to graze – leaving a long rope attached to his head collar because I wasn't convinced that I'd be able to get hold of him again otherwise.

With the equines all happily settled for the night, we headed to Mum's house, where good food, hot showers, a washing machine, and a comfy bed awaited.

28

East Clare

After two days of well-deserved rest and relaxation, Vlad flew back to England to work for a week leaving me to tend to the horses, make a start on Micheál's training, and settle briefly into life in County Clare.

On the surface Clare seems like a sleepy, nowhere sort of a place in the back of beyond, where nothing exciting ever happens. If you look carefully, however, you'll soon find that it is actually a hive of quiet activity, teeming with a whole host of interesting people. There are healers and herbalists; homeopaths and white witches; psychics and story-tellers; biodynamic farmers and anthoposophists; yoga centres and alternative communities; and off-grid dwellers living self-sufficiently in eco-houses – all quietly getting on with their lives, tucked away here in the westernmost reaches of Europe. Amongst them, I found Ellen.

A friend of my mother's had said I should go and visit Ellen because she was a really interesting person and 'a bit of a gypsy' – whatever that meant. Keen to find out, I rang her up, borrowed mum's car, and headed off into the wilds of County Clare, following some carefully given directions.

After a short drive through undulating green countryside, I found myself in front of a small, single-storey cottage. It was tucked away in the middle of an overgrown garden full of herbs and potted plants, and half-hidden under thick creepers. Beneath a tree in one corner of the garden stood an old bowtop wagon.

Ellen came out to greet me and invited me into her cosy cottage for a light lunch. She was in her fifties, with short blond hair and a round, pleasant face. She was a beautiful soul, Ellen. She was a wisewoman and a healer; intuitive, empathic and a real free-spirit – but she was grounded, too. Although English by birth, Ellen had lived in the west of Ireland for many years.

After we'd eaten, we went out into the garden and sat down by the firepit where a stack of logs were smouldering away. We sipped cups of herb tea while Ellen told me tales of Ireland before it became modernised not so very long ago. She had spent years hitch-hiking around the country, living from one meal to the next and sleeping rough under the hedges. She'd also spent time travelling between the fairs in bowtop wagons – dealing in horses and selling her homemade healing creams to the Travellers.

She handed me a pile of old black and white photographs while she talked. They were memories of her days on the road. Bohemian women in long, flowing dresses, and unshaven men wearing trilby hats, walked alongside hairy Gypsy cobs pulling bowtop wagons; on the driver's seat sat scruffy-looking children with unkempt hair. The images were of a time before motor vehicles had become commonplace throughout the west of Ireland; before Celtic Tiger had brought economic growth and rapid modernisation to rural Ireland; and before all the Travellers' pull-in places had been completely blocked off by local councils. These were memories of a simpler, freer life.

Thus the afternoon drifted by as I soaked up Ellen's stories, watching flames dance in the glowing embers of the wood fire, caught up in that not-so-distant but somehow magical past.

When we parted, Ellen gave me a pot of her home-made cream. She said it would heal any cuts or sores that either we or the horses might develop on our onward journey.

East Clare

Another person I encountered during my stay in Clare was Ruth. She was a writer and a story-teller who had arrived in Ireland from her native Scotland and had put down roots in East Clare. Someone we'd passed on the road to Mountshannon had pointed us her way, but at the time I'd had my hands full with the mule so I'd taken her phone number and arranged to meet her a few days later.

I pulled into the driveway of her little cottage tucked away in the gently rolling hills behind Bodyke. The place was comfortable and homely. There was a pile of freshly harvested apples on a table outside the kitchen door, and a young rowan tree grew beside the house. Ruth said it was customary in her family – always had been – to plant a rowan tree by the door to ward off evil spirits.

Ruth was housebound with a broken ankle, so she was more than happy to sit and talk – telling me some tales and filling me in on a bit of local folklore. It was from her that I first heard of Biddy Early, a famous wisewoman who had lived in East Clare during the 19th century.

Biddy Early was renowned throughout the west of Ireland for her remarkable healing powers, and people would come to her from all over the country in search of a cure for their ailments. Biddy never took payment for her work; instead people brought her gifts – whatever they could manage – be it eggs, butter, a loaf of bread, or even a bottle of Poitín. In this way Biddy would tend to even the poorest of peasants who could not afford the services of a doctor.

Along with her healing powers, Biddy also possessed the Sight and she knew who was coming to see her and why, before they even arrived. It is said that she got this power from a blue glass bottle which she carried with her everywhere. If she wanted to know something, she would simply raise the bottle to her eye, and look inside it. Ruth told me the story of how Biddy came by this magic bottle, as we sat outside her cottage drinking coffee and soaking up the late summer sunshine.

Biddy had a grown-up son named Tommy who used to love dancing and playing cards, and he was a great hurler as well. One night Tommy went over to the neighbour's house for a game of cards. After the game had finished he set off for home. It was a full moon that night, so there was plenty of light for him to see his way as he walked back along the lanes. Suddenly he heard laughing and shouting from the field beside the road. Peering over the wall, he saw a crowd of little men playing hurling. These were the Good People, the fairies. When they saw Tommy, they called over to him: 'Come on and join us, we're one man short. You're a good hurler aren't you?'

Tommy leapt over the wall, joined in the game, and the side he was playing for won the match. The little men picked Tommy up and carried him around on their shoulders, cheering with delight. They set him down again, clapped him on the back, and when it was time for him to go, they handed him a blue glass bottle.

'What am I supposed to do with that?' Tommy asked, curiously.

'Take it home with you and give it to your mother. She'll know exactly what to do with it,' they replied. So Tommy took the bottle, stuffed it inside his shirt, and continued on his way.

In the morning, whilst he was sat having his breakfast, he suddenly remembered the bottle.

'Oh, here mother, something happened last night,' he said, and told Biddy the story before handing her the bottle. 'They said you'd know what to do with it.'

Biddy took the bottle, looked into it, and there she saw the future. From that day forward she used this gift from the fairy-folk to help people and heal the sick. When she was very old and on her deathbed, Biddy gave the bottle to a priest with instructions to throw it far out into the lough by the house so that the bottle might never be used for evil.

About fifty years ago, Ruth said, the lake was drained, and although they found many glass bottles at the bottom of it, there wasn't a blue one among them.

While we waited for Vlad to return, Micheál's training also began in earnest. I sallied forth to the field each morning bearing offerings of apples and carrots, and spent time trying to catch, groom, and lead him around.

Some days he would be co-operative and eager to please and nothing I did would seem to bother him. All the grooming, petting, and tying up – although clearly a novelty to the little mule – were tolerated with an incredible amount of calm patience. On those days I'd leave the field feeling happy and confident that he would be a wonderful addition to our little fellowship, and that the onward journey would surely be a success. On other days, however, Micheál would flatly refuse to be caught, choosing instead to run tight circles around me – always just out of reach – and keeping a close watch on me with his one beady little eye. If I got too close to him, he would threaten me with a kick. On those days I found myself questioning my wisdom in acquiring this grumpy, half-feral creature, and I wondered uncertainly whether it would be better to find him a nice home somewhere nearby where he could live out his days in relative laziness. But the next time I went to see him, I'd find him agreeable and willing to work with me again and all those nagging doubts would vanish.

I very quickly came to realise that here was an animal who could not be forced into anything against his will. Bullying and coercion – which are not my cup of tea, anyway – were absolutely out of the question with this little mule. If I wanted to work with him, then I'd have to gain his absolute trust – but it's a big ask when you have just two weeks to do it!

After the first few doubt-filled days, Micheál began to relax. His refusals to be caught were less frequent and he became gradually more co-operative. After a while he even started to seek out attention. Although he still viewed me with an air of suspicious uncertainty, he would do whatever I asked of him without argument, and that was as good a start as I could have wished for.

I would spend half an hour each morning and evening leading him around the paddock, touching him all over with a long stick, and tying him up to groom him. It wasn't long before I even managed to put a saddle pad on his back – which he tolerated admirably. If Micheál was going to join our little adventure, then he'd have to carry a pack. On that point I was adamant.

With that in mind, I set about making him a special pack-pad from an old hessian sack. I stitched the sack into carefully measured sections, which I stuffed tightly with straw to make two hard, wide panels in the middle. These gave good spinal clearance and helped to distribute the weight of the packs along his back. The sides of the blanket I also filled with straw and quilted to make a thick layer which would protect his ribs from rubbing. I found some old leather belts in a charity shop and re-purposed them as a breast collar, breeching, and girth straps, and I bought two small rucksacks which I could fix onto the pad with lengths of thin rope. It was a crude but functional set up and, to my great surprise, Micheál accepted it without complaint.

The horses were gradually becoming accustomed to their new long-eared companion, although Oisín was still very protective of Dakota and tried to herd him away from the fence whenever he came too close to Micheál. On several occasions, when Oisín wasn't looking, I caught Dakota gently licking Micheál's face and grooming with him over the electric fence. Dakota is much more kind and easy-going by nature than his jealous friend.

After a week, when I was certain that I could catch the mule if I needed to, and that Oisín had given up on any seriously murderous intentions, I removed Micheál's rope and let them all run together. In this way the lazy days drifted slowly by while I prepared Micheál for life on the road.

29

Back on the Road

Autumn arrived on the third of September. I knew it the moment I stepped out of the door that morning. The sky was clear and blue, the sun shining, but there was a chill to the air and a distinct change in the quality of the light, which told me that the long, languid days of summer were over and autumn had begun. Blackberries clustered on the bramble thickets that lined the lanes, and the hawthorn and rowan trees growing in the banks were laden with bright red and orange berries; the green canopies of the trees were tinged with yellows and browns as the leaves began to fade; and the nights were starting to draw in – a pressing reminder that the countdown to winter had truly begun. We needed to push on if we were to reach the end of our journey before the cold, wet weather set in.

Vlad returned on the fourth of September and a few days later the vet arrived to microchip the strongly objecting Micheál. It wasn't until the eighth of September that we finally set off again.

The day dawned grey and overcast; a cold, steady drizzle was falling as we caught the horses and Micheál, tacked them up, and headed off into the Slieve Aughty Mountains. We followed a gently climbing tarmacked road which eventually gave way to a rough, stony track that wound over the empty hills and then dropped down through large swathes of forestry plantation towards Lough Graney.

I rode for a few miles, leading Micheál off Dakota because Oisín refused to have the mule walk behind him and tried to kick him if he got too close. Micheál was wary even of gentle, good-natured Dakota and so he hung back on his rope the entire time, pulling my arm half out of its socket. Frustrated, I soon gave up trying to lead him like that so I dismounted and, handing Dakota's rope to Vlad, I walked with Micheál instead. The mule seemed happier then, and tripped merrily along beside me through the wet hills and dark forests.

The day was still; not a branch stirred, and not a bird sang in the forest where nothing grew. In the middle of the afternoon the rain eased off, but mist still trailed through the treetops and the only sound was of the horses' hoof-beats along the stony tracks.

It was a short ten-mile journey to our stop – a smallholding near Flagmount where Mum's friends Agnes and Mike had offered the horses a place for the night with their herd of milking goats. By the time we arrived my feet and legs were aching, and Micheál was tired and grumpy after his long walk through the mountains. It was a relief to unclip his lead-rope and turn him away amongst the goats in their large field full of rich grass. All things considered, he'd been very good.

Mum arrived, and after a cup of tea and a chat with Agnes and Mike, we piled into the car and headed back to Mountshannon. A last hot shower was had, good food enjoyed, and our packing for the morning completed, before finally curling up in a comfortable bed. Who knew when we'd be able to enjoy any of those luxuries again!

In the morning Mum dropped us back over to the horses. We said our goodbyes then headed down to the field. Dakota and Oisín were keen to be on the move again and came straight up to the gate to be caught. Micheál, however, had had enough. He didn't like travelling, he'd decided; life on the road was not for him and there was no way he was going to spend another day traipsing about the countryside, thank you very much! He communicated these sentiments by emphatically refusing to be caught.

I had made the rather stupid mistake of removing his lead-rope the night before – certain that I'd be able to catch him again, just as I'd been doing for the last few days in the field in Mountshannon.

I could not have been more wrong.

At first I tried gently coaxing him with apples, which he graciously accepted, but then he'd shoot off before I could get hold of his head collar. Next he trotted tight little circles around me – always just out of reach – before heading off to the far side of the large field, where he'd stop and stare at me with his one good eye, flicking his tail in angry defiance.

I tried cornering him between the fence and a ditch but, after careful consideration, he leapt gracefully over the ditch and trotted away across the field.

When that failed, I tried to trap him in between clumps of brambles and fallen trees. I soon discovered – to my dismay – that not only was he a strong-willed, determined little beast, but that he was also incredibly agile. The fact that he only had one eye did nothing to deter him from picking his way through the most impenetrable of thickets, and clambering over or squeezing under the dense piles of dead branches and fallen trees – and all this without sustaining so much as a scrape. Even electric fencing didn't worry him enough to stop him pushing underneath it or barging through it. Nothing, it seemed, could stand in Micheál's way when he set his mind on something, and his mind that morning was fixed firmly on one goal: to evade capture.

Round and round the field we went, and every attempt I made to catch him was skilfully dodged. Twenty minutes passed; half an hour; an hour; two hours – I did everything I could think of to get hold of him that morning, and still the obstinate little devil would not give in.

I cursed that mule quietly to myself, swore at him vehemently under my breath, and called him every name under the sun – trying all the while to maintain an outward cool. If I let my irritation and frustration show, it would only make the situation worse.

In the meantime Vlad had taken the horses up to the house where he'd groomed and saddled them, loaded up all the gear, and was waiting patiently for me to arrive with the mule. Eventually he decided to come and see what was taking us so long.

We joined forces trying to outsmart Micheál and block his attempts to avoid entrapment – but to no avail. The savvy little mule was just too intelligent and always one step ahead of us. The whole thing had now become a terrific game and a battle of wits, in which the score was something like 50-0 to Micheál.

Fed up of trying to coax the mule nicely into submission, Vlad suggested we try herding him into a length of rope. We tied one end of a rope to the fence and Vlad held the other. Micheál obligingly trotted into it; then Vlad launched himself at the creature in a vain attempt to hold him long enough for me to clip the rope onto his head collar.

This clumsy attempt at a rugby-tackle was an unmitigated disaster.

Micheál panicked; he wriggled, squirmed, and kicked violently as he fought his way through the rope – breaking the fence, and leaving Vlad lying prostrate in the mud – and made off across the field again, more determined than ever to not be caught.

It was three long hours of this ridiculous and frustrating circus before we finally managed to get hold of him. In the end, this feat was achieved by dropping a large noose over his head from a safe distance by means of a long stick. Once the rope was around his neck, the mule gave up the fight and let me catch him as though it were the most natural thing in the world. I honestly could have killed him that day, and I vowed never again to take his rope off!

30

Micheál's First Days on the Road

It was early afternoon when we finally set off from Agnes's in the direction of Flagmount, skirting the shores of Lough Graney.

There is a Clare folktale that tells of a chieftain in ancient times who lived in the Slieve Aughty Mountains. This chieftain had a daughter named Grian who was so beautiful that many young men wanted to marry her, but in order to do so they needed to know who her mother was.

Curious, Grian asked her father one day. The chieftain said that her mother had been no mortal woman at all, but a beautiful sunbeam. Grian was so distraught by this news that she threw herself into the waters of the lough and drowned. Her body floated down the river into Lough Derg and washed up in an oak wood near Scarriff. It was taken from there and buried under a mound nearby. All those places now bear her name: Lough Graney, the River Graney, Derrygraney, Tuamgraney – Tomb of Grian, or Tomb of the Sun.

I was in no mood for folklore or fairy tales that day, however, because once again Micheál was protesting at having to walk, refusing to lead off Dakota, and – after the rugby-tackling incident that morning – wanted nothing to do with Vlad.

If Vlad tried to lead him Micheál would dig his heels in, hang back on the rope, and refuse to budge. For a tiny little creature, he could certainly pull off a convincing impression of an anchor, and so he would stand rooted firmly to the spot until I came to walk with him once more. Had I not been so tired and fed up from the morning's antics, I might almost have been flattered.

Mules are pretty unforgiving creatures and they're well known for holding life-long grudges against people who have wronged them or betrayed their trust. Micheál had certainly taken offence at Vlad's rough handling of him and had developed a deep dislike for him. His feelings of animosity and resentment were fully reciprocated by Vlad, who had gone right off our little mule since that morning, too. It was as much as I could do to stop Vlad from opening the nearest field gate, shoving Micheál inside, and riding off without so much as a backward glance.

We covered fourteen miles following narrow roads that dipped and climbed their way across low hills. Sometimes we trudged through acres of barren forestry, other times over open heathland where purple heather and indigo-blue devil's-bit scabious grew thick along the verges, all interspersed with yellow flowering gorse. In the distance we could just make out silver Lough Cutra, and beyond it, through a haze of falling rain and sunbeams, loomed the silhouette of the grey Burren.

Just as the light was starting to fade, we happened upon a kind man who invited us to camp in the large, overgrown field behind his house. We tied the horses and Micheál to a rickety fence and started to untack them while our host went to sort out some water for them all.

I was so preoccupied trying to remove Micheál's pack without getting kicked (because one again he was tired and very grumpy after a long day's walk), that I didn't notice our host stick the end of a hosepipe into a bucket right under Dakota's nose. Havoc ensued when he suddenly turned the hose on full blast, startling poor Dakota who leapt several feet into the air, bucking wildly and sending his packs flying in all directions before falling over and getting his legs caught in the rails of the wooden fence. When he got to his feet again I managed to calm him down, and he stood snorting and quivering with fright while I whipped off his remaining packs. To my great relief both he and the packs came out of the ordeal unscathed.

Micheál's First Days on the Road

We let the horses loose to graze, but Micheál, we decided, would be better off tethered to avoid a repeat performance of the morning's shenanigans.

At first our little mule was agreeable to this decision and I kept a careful watch on him while we set up the tent a few yards away. After an hour or two, however, Micheál became annoyed at his restricted freedom and began kicking furiously at his rope in a determined effort to break it. His kicks were well calculated and perfectly aimed!

I hastily constructed a ring of electric fencing around him so he couldn't pull the rope taut to attack it. This worked for a while, but around two o'clock in the morning I was awoken by a commotion and got up to find Micheál on the far side of the electric fence, once again taking carefully aimed shots at the rope with his hind feet.

I swore at him a lot then, and needless to say we didn't get much sleep that night as he continued to push his way through the electric fence at regular intervals. To make matters worse, in the morning the heavens opened and it bucketed with rain.

We waited an hour, hoping the rain would ease off, but after a while we decided we'd better suck it up and hit the road. In no time at all we were soaked to the bone and feeling fed up and miserable. Once again Micheál refused to lead from the horses, so I had to walk with him, and for the hundredth time we found ourselves wondering whether acquiring the mule really had been such a good idea after all. So far he was not taking to life on the road with any enthusiasm. Actually he seemed to hate it. Was it fair to drag him along for hundreds of miles if he wasn't enjoying the experience? I was beginning to think not. Perhaps we could leave him somewhere until we finished the journey and pick him up on the way back, I thought in desperation.

We made it all of a mile along the road towards Peterswell before a woman in a passing car pulled over to ask what we were doing. We must have looked as fed up and miserable as we felt, because she lost no time in inviting us back to her place for a cup of tea. Vlad wasn't keen on the idea because it meant backtracking, but I was. After a brief discussion we 'compromised', and retraced our steps for three miles into the mountains back the way we'd come.

The kind woman was called Suzanne. She and her husband, Matthew, were both English and they lived with their two young children half way up a mountain in a quirky stone cottage, which was painted a startling shade of pink. Matthew dealt in old cars and antiques, and Suzanne was a talented textile artist who ran a successful online business making creams and lotions, and painting tiles.

A German woman named Sabine lived on their land, too. Matthew said she'd arrived one day to buy a caravan, asked for a place to stay – temporarily he'd assumed – and had then built herself a nice little hut at the end of their field and never left. Sabine spent most of her time hand-spinning sheep fleeces and going round the local markets selling her knitted hats and spools of homespun, home-dyed wool.

Just up from the house was a large field which contained a lone geriatric goat. Matthew and Suzanne said we could put the equines there to graze. They also said we could leave Micheál behind as company for the goat if we liked. We could pick him up whenever we were ready.

The offer was a tempting one, but all the same, I felt a pang of remorse. I had made a commitment to this stubborn little creature and, as difficult as he was, I didn't want to give up on him so easily. Vlad, on the other hand, was all for leaving Micheál behind, and I suspect that if we could have asked the horses, Oisín would have agreed. In the end I decided that it would have to be Micheál's choice.

'Look, if you want to come with us, then you'll have to let me catch you in the morning, otherwise you're staying here!' I said to him firmly, before dropping his rope and watching as he stalked off across the field to graze.

In the morning, after coffee and breakfast with our hosts, we packed down our things and went to fetch the horses. Micheál kept his wary distance, clearly not wishing to be caught. I tried coaxing him with apples, but he remained resolutely aloof so I left him to it, loaded up Dakota and just before we were about to leave, I gave Micheál one final chance.

'This is it,' I said to him. 'You either let me catch you, or you stay here with the goat. It's your choice!'

Micheál's First Days on the Road

To my utter astonishment, and great relief, he let me get hold of the rope and stood quietly while I put his blanket on and loaded up all the packs. He'd made his decision. He was coming too!

31

Galway

The rain eased off as we came down from the mountain and made our way into Peterswell. There wasn't a great deal in Peterswell. In fact the pub and the scant smattering of houses there barely constituted a village. I was just coming to the conclusion that there was absolutely nothing of interest here at all when, on a bend in the road ahead, I suddenly noticed a little cottage. The small, single-storey building was all boarded up and falling into disrepair, but on the wall facing the road was a sign that read 'Joe Cooley's House.'

Joe Cooley was a famous accordion player who had been born and raised here in Peterswell in the early 20th century. His musical career had taken him all over Ireland, England, and the United States, and he is still regarded as one of the most accomplished accordionists of his time. His playing inspired many people and his tunes are famous the world over. In fact one of the first tunes I learnt to play was Cooley's Reel. Although it is often argued whether or not Cooley actually composed the tune, it was certainly he who made it popular. To this day, it is one of my favourite melodies.

Sadly, there was little more to see here than the boarded up house and the sign. It was a poor memorial to such an influential musician.

We traipsed on along the little lanes bordered by dry-stone walls which were as much empty space as they were stone; the rocks of these peculiar structures were balanced precariously one on top of the other, leaving large gaps between them. They looked as though they would topple over with the first strong gust of wind. I couldn't decide whether they had been carefully constructed to look like this, or whether they had simply been thrown together in a careless, haphazard manner.

The ruins of old cottages with ivy-covered trees and bramble thickets poking through half-collapsed roofs were a regular feature here, too. No efforts had been made to salvage or rebuild these old houses, or even to demolish them. They stood neglected and forgotten – left to disintegrate into a pile of rubble while a brand new, much bigger house was built alongside.

It was nearly seven o'clock when we arrived, footsore and weary, in Carrabane, a small village not far from Athenry. The evening was drawing on and we still hadn't found anywhere to stop.

We tried to flag down passing cars to ask, but most people ignored us and sped on by. The few who did stop looked at us warily and said they didn't know of anywhere we could camp. In the end I headed up a long drive between acres of lush grass to ask at a farm for a field. The woman who answered the door gave me a look of deep suspicion and told me they had no fields, and they knew of nothing nearby. I wasn't convinced, but I thanked her anyway and headed back to where I'd left Vlad in charge of the animals.

Micheál had somehow managed to get himself tangled up in his rope and his pack had slipped, too. When Vlad tried to help him, Micheál had threatened him with a kick, so they'd both had to wait patiently for my return to disentangle the mule and right his packs again.

We were just beginning to despair of finding a stop that night, and had started to look about for a field that we could sneak into, when I finally managed to flag down a passing pick-up truck. The man driving it was a cattle farmer and, after chatting with us for a while, he eventually invited us to camp at his farm about three miles down the road back the way we'd come. Although Vlad hated backtracking and argued that we should push on and keep looking, it was too late in the day to risk it, so back we went.

Darkness was falling when we arrived at the farm. The nice farmer, Paul, met us and said we could sleep in the large open-fronted barn and the horses could have the small paddock behind it.

The barn had running water and electricity, and on an open platform under the roof we found a mattress. Once dusted off and covered with the woollen blankets we were using beneath our saddles, it made a very comfortable place to sleep.

The next day we skirted Athenry, crossing over the M6 motorway which ran from Dublin to Galway. The countryside was flat and dull for most of the journey, until we picked up a pleasant road which skirted the edge of Lough Tee Bog Nature Reserve. Here the landscape became prettier, with rolling green pastures and grey stone walls set against a backdrop of dark evergreen forests; native trees grew dotted about the large pastures, or in long lines where old hedgerows had been neglected and left to their own devices. Overgrown hedges were a feature throughout the Irish landscape and it was a rare thing indeed to see one well maintained.

Once again that evening we struggled to find anywhere to stop. Everyone seemed wary and mistrustful of us and most people we asked shook their heads and said there was nowhere for us to camp – even though we were surrounded by acres of empty fields. One man at whose farm we had stopped to enquire, and who had sent us packing in a rather hostile manner, passed us a mile down the road in his car, took a picture of us on his phone, then turned around and headed back the way he'd come. I half expected the police to arrive then and ask us what we were doing.

The uncharacteristic suspicion and hostility with which we were suddenly confronted was – we eventually discovered – because the famous horse fair at nearby Ballinasloe was coming up in a few weeks' time. Thinking we were Travellers, people were wary of us and gave us a wide berth.

After several more rejections and blank stares accompanied by a shrug of the shoulders, we finally found a kind farmer near Skehana who offered the horses an enormous, heavily fertilised pasture. It wasn't good grazing for them, but we couldn't risk pushing on in case we didn't find anything else.

Vlad pitched our tent in the hay barn amongst the big round bales. It was well sheltered from the wind which was blowing fiercely and brought with it regular rain showers. The thick layer of old hay covering the floor of the barn beneath the tent made a nice, soft bed and we slept soundly.

In the morning, the farmer's elderly mother, Noreen, invited us in for a breakfast of bread, jam, and strong black tea. She welcomed us like old friends and it was late before we finally got away and headed on into Skehana.

Skehana was a bit of an empty place with a smattering of houses and not a lot else, but we had it on good authority that the area was full of buried treasure – crocks of gold to be precise, most of which could be found beneath whitethorn trees guarded by cats. In fact the name Skehana is actually derived from the Irish Sceith eánach meaning place of the whitethorn.

Although we saw many thorn bushes in the fields and hedgerows none were especially conspicuous, nor were any guarded by cats, and we saw no hazel woods either – in which one magician had supposedly buried his treasure. It was a little disappointing because – given the sorry state of my bank account – a crock of gold would not have gone amiss.

The whitethorn tree in Irish folklore is also beloved of the fairy folk, and the doings of the Good People abounded in the local folktales. One story tells how a farmer from Skehana had been so unlucky with crops and livestock that he'd travelled all the way to East Clare to seek out wise-woman Biddy Early for advice. After looking into her blue glass bottle, Biddy told the man that he should stop throwing old food and kitchen scraps out of his back door, because in doing so he was throwing them onto the fairies' dinner table. This caused them such offence that they had given him bad luck in punishment. The farmer returned home and did as Biddy said. He never threw anything out of his back door again and from that day forward he prospered.

Our route took us along quiet lanes over flat, sparsely inhabited countryside; acres of open pasture eventually gave way to large areas of forestry plantation and small patches of scrub. Intermittent but heavy rain showers plagued us throughout the day, a cold wind blew in our faces, and we prayed for somewhere dry to pitch the tent that night.

When we reached Mountbellew we began asking around for somewhere to stop. The people here were a little more friendly than those we'd encountered over the past few days.

'Try Kevin Higgin,' everyone said. 'He's sure to help ye's out.'

After some traipsing up and down a hair-raisingly fast main road, we finally found Kevin Higgin and his lovely partner, Geraldine. They offered the horses a nice paddock next to their little coloured cob, and they let us make camp in the barn – safe from the pouring rain.

Blankets and inflatable mats were thrown across carefully arranged hay bales to make one of the most comfortable beds we had slept in for weeks!

The weather was becoming wetter by the day, the evenings were drawing in, and temperatures were in steady decline. There was no doubt that autumn was well and truly upon us. Nights in the tent were growing uncomfortably cold, so along with finding good grazing for the horses we also tried to find places that had a barn in which we could pitch the tent. This not only ensured all our equipment stayed dry, but it gave us shelter from the wind and kept the temperatures at a near-bearable level, too. It was a far cry from the scorching heat we'd set off in, nearly twelve weeks earlier. That was now little more than a distant but happy memory.

Micheál was finally coming round to the idea of walking every day and carrying light packs. In spite of this, we couldn't persuade him to lead from the horses, so Dakota became the full-time packhorse and one of us had to walk with the mule at all times. Vlad and Micheál still harboured a deep resentment towards each other, and both flatly refused to reconcile their differences. Vlad strongly objected to leading Micheál, and Micheál strongly objected to being led by Vlad, so as a result I was forced to lead him day in, day out, for many miles. While I didn't particularly mind the exercise, it was a blessing indeed when

Vlad and Micheál agreed to cooperate for a mile or two so I could rest my legs and ride Oisín.

Unfortunately Vlad and Micheál's wary truces were always short-lived, and were usually broken by Micheál trying to kick Vlad – at least that's what Vlad said. Since Micheál couldn't say anything in his defence, and I never saw it happen because I was riding in front, I had to accept Vlad's side of the story, dismount, and take over leading the dear little creature once more.

Because the days were getting rapidly shorter, and both Micheál and I grew tired after several hours of walking, our daily progress became much slower than before. Instead of covering twenty-odd miles a day – as we'd been doing up until the mule joined our fellowship – we were now barely managing more than twelve. Whenever Micheál became tired and hungry, he would also become incredibly grumpy. He'd dig his heels in obstinately, pull faces, and refuse to budge unless it was to go and graze the nice grass which grew thick along the verges. He was a rather opinionated little being.

Unsurprisingly, Irish folklore is full of tales about mules and jennets and their many misdeeds – jennets in particular, I might add. They were renowned for being cantankerous little creatures and it was even said that they are cursed because one refused to carry the Virgin Mary. Not only did it refuse to carry her, but it kicked her, too.

After a few weeks in Micheál's company I had no trouble believing it!

32

Roscommon

By the time we reached Creggs, we were more than eight hundred miles into the journey and, for the fourth time on the adventure, Dakota needed new shoes.

We had spent a pleasant, dry day following little roads over bogs and pasture and through acres of thick forestry. Often we passed fields of grazing horses, and Micheál liked nothing better than to stop and talk to them. Most of the time the horses, who had come galloping over to the fence to get a closer look at us, would turn tail and bolt the moment Micheál opened his mouth. Was it any wonder the poor creature was so grumpy if that's how he was received by every equine he encountered?

The quiet little lanes eventually brought us onto a main road, but it wasn't busy and we followed it happily for several miles. I was riding Oisín, Vlad and Micheál were cooperating well for a change, and all was right with the world.

As we came to the outskirts of Creggs, we saw a cyclist approaching on the opposite side of the road. The horses had seen hundreds of cyclists on the journey so far and there was nothing especially different about this one that I could see. What inspired Oisín to suddenly take fright, spin a hundred and eighty degrees, and bolt off up the road in terror is anyone's guess, but bolt he did – with Dakota right behind him.

Dakota's shoes by this time were worn flat and had no grip at all so he skidded wildly, leaving a nice set of marks in the asphalt as he dodged a car, narrowly avoiding a head-on collision. After a short, but energetic sprint, I managed to stop Oisín. Dakota stopped too, and I retrieved his rope – which I'd led go of in all the mayhem. Both horses nickered softly to each other in an endearing display of mutual concern before we turned around and headed back to where Vlad, Spirit, and Micheál were standing quietly waiting for us. In spite of the horses' mad panic, Micheál hadn't batted an eyelid because – unlike them – he had a very sensible approach to life and he thought things through before he acted.

When we finally arrived in the town, slightly flustered but all in one piece, we tied the animals up to some railings outside the school. I kept watch over them and chatted to some curious locals while Vlad went to stock up our supplies at the petrol station a little ways up the road. He returned with half the shop and the phone number of a man named Brendan who owned some stables on the other side of town.

Brendan wasn't home, but his lovely young partner, Margaret, was. She didn't hesitate in offering our four legged friends a small paddock beside the large, immaculate yard where she and Brendan bred, broke, and trained horses. A barrow-load of sweet smelling haylage was given to our eager equine companions and then Vlad, Spirit, and I made camp in an empty stable block. Once we were settled in, Margaret invited us into the house for tea and toast, and she kindly arranged for her farrier to come out and shoe Dakota the following day.

The farrier arrived in the late morning and Dakota was brought out and tied up on the yard. Perhaps it was because Dakota hadn't been getting nearly so much exercise over the last few weeks and was stuffed full of rich autumn grass, or perhaps Patrice was a bit too brusque for Dakota's liking; whatever it was, this farrier – Dakota decided – was not a nice one. He scooted about on the end of his rope with his eyes protruding wildly, firmly declining to give his feet.

Meanwhile Oisín was kicking up a fuss in the paddock nearby. Ever-protective of his friend, he was concerned for Dakota's well-being and worried that Dakota was receiving food that he wasn't.

Margaret took hold of Dakota's rope while I went to fetch Oisín in and stop him from destroying all the fencing. Margaret soothed Dakota, tapping her fingers lightly along his cheek and talking quietly to him until he relaxed. I watched, fascinated. At last – half-hypnotised by her gentle touch – Dakota reluctantly let the farrier get to work.

Once the shoes were on we packed up, thanked our hosts, and hit the road, following lanes through pleasantly undulating countryside, past a smattering of houses and small farms. We crossed the River Suck near Dunamon Castle – one of the oldest inhabited buildings in Ireland – and continued on across flat countryside. The tumbled-down ruins of an old castle stood on the riverbank here, and fairy forts could be seen in some of the surrounding fields, their untouched trees growing in tight circles on raised banks of earth.

It was now eight days since we'd left Mountshannon and eight days since we'd last had a shower or washed our clothes. All those nights sleeping in dusty hay barns and days spent traipsing about in sweaty waterproofs had taken their toll. To say that we were none too easy on the eyes and none too easy on the nose is an understatement. Having a wash and getting our clothes clean was fast becoming a priority.

Several friendly people had stopped to offer us a place to stay, but it had been too early in the day and we'd only covered a few miles, so we pushed on. It wasn't until we reached Ballymacurly that we began looking for a camp. Although houses were frequent and most had fields alongside, we hardly saw a soul to speak to and the one farmer we did see merely muttered something inaudible and walked off. We took that to be no. An Englishman outside a house gave us the number of a local man who owned most of the land in those parts. I rang the number, but the man was half-deaf, the line was poor, and he couldn't hear a word I said so I gave up. Eventually a woman in a passing car told us to try at the Old School House about a quarter of a mile further along the road. The people there were friendly and might help us out, she said.

After a few minutes we drew level with a large, gloomy old building of grey stone. The tall bell tower in the middle of the slate roof over a large gable told us that this was the place we were looking for.

Beside the rather imposing house was a field of lush grass, and a young couple were making their way up from a polytunnel which stood at the far end of the field. We called out to them as they approached and they came over to the gate.

'Sure, work away,' they said when we asked if we could camp. 'There's no shortage of grass!'

Electric fencing was hastily put up, the equines were unloaded and turned away to gorge on rich grass, then we introduced ourselves properly to our hosts.

Broc and Edel were a lovely couple of young newlyweds. Edel worked as an interior designer, and Broc worked on a large-scale dairy farm. They'd both done a lot of travelling so they were sympathetic to our needs.

Showers were offered and gratefully accepted, the washing machine was put at our disposal and filled, and our hosts even offered us their spare room for the night – but no dogs were allowed in the house so Spirit would have to sleep outside. I knew Spirit wouldn't be happy with that arrangement and would cry the entire night long, so we pitched the tent in a sheltered corner of the front lawn and settled in.

Our hosts had converted the interior of the old schoolhouse very tastefully and in keeping with the building. The back door opened into a utility room and a long flagstone corridor led off it, which ran from one end of the house to the other beside the main schoolroom. At the far end of the corridor were two bedrooms and a quirky little bathroom, and the main schoolroom had been turned into a large, open-plan living room, kitchen, and dining area with a beautiful mezzanine platform over the kitchen. The high vaulted ceiling gave the space some nice acoustics and, after we'd showered and settled in, I fetched my fiddle and played a few tunes.

Broc and Edel were pleasant, easy-going people, and both possessed a good sense of humour. Edel especially had a knack for telling a story and she entertained us with hilarious tales of a local character named Paddy. He was a bit of a rogue and was forever having run-ins with the law, but he only ever tried to get one over on 'the man' and never on his neighbours. He had a great way with words, Edel said, recalling how Paddy had once asked her mother for a dance at a local function.

'Ah Paddy, would you stop your messing!' Her mother had insisted, when Paddy's hands had started to wander.

'Well aren't you awful giddy for one who was so well handled,' he'd replied – referring to the fact that Edel's mother was a good Catholic who'd had many children.

In the morning the couple invited us to stay another night. It was an offer we couldn't refuse because we'd not had a day off since leaving Mountshannon. While the horses and Micheál rested and did their very best to clear the overgrown field, Vlad and I spent the day helping our hosts shovel cow manure into the polytunnel. It seemed like a fair exchange for their hospitality, but we regretted washing our clothes the night before because now they were smellier than ever!

We were approaching Tulsk now, a modern town located near the ancient site of Cruachan, capital of Connacht, and seat of the legendary warrior Queen Medb.

One of the best-known Irish sagas – Táin Bó Cúailnge, or the Cattle Raid of Cooley – tells how Medb once tried to steal a bull from the neighbouring province of Ulster.

Medb at the time was married to Ailill and one day the two were comparing their wealth. They found that they were equal in all things, save one: Ailill possessed a mighty stud bull – the white horned Finnbhennach – and Medb had none to equal it. The Finnbhennach had actually been born into Medb's herd of cattle but, not wishing to be owned by a woman, he had gone to join her husband's herd instead. This was a point of great grievance for Medb so, in order to restore equality between herself and her husband, she resolved to steal the Donn Cuailnge – the Brown Bull of Cooley – who lived on the Cooley Peninsula in Ulster.

Word got out of Medb's plan to steal the bull, so the Ulstermen gathered an army to stop her, and Medb gathered an army to meet them. Just days before the battle the men of Ulster fell suddenly ill and not one of them could fight. Only the young hero Cú Chullain, son of Lugh, was able to defend Ulster against the attack.

Invoking the ancient right to single combat at a ford, Cú Chullain managed to stave off Medb's invasion. Each day Medb would sent a champion to the ford to do battle against Cú Chullain and each day the river ran red with the blood of Connacht men as Cú Chullain won the fight.

Sometimes the Tuatha De Dannan helped Cú Chullain in his battles, and at other times they hindered him. Thus things continued for many months and Cú Chullain remained undefeated.

In desperation, Medb finally incited CúChullain's own foster-brother Ferdiad to fight Cú Chullain at the ford. The battle raged for three long days because the two were brothers and did not wish to kill one another, but on the third day Cú Chullain reluctantly slew Ferdiad. Cú Chullain was wounded so badly in the fight that he was forced to retreat in order to rest and heal. By that time the men of Ulster had recovered from their illness and they arrived to join the battle.

The Ulster men defeated Medb's armies, but in the meantime she had found the Brown Bull of Cooley, so when she retreated she took him back to Connacht along with fifty of his heifers.

What Medb did not know, however, was that the Brown Bull of Cooley and the Finnbhennach were sworn enemies so they fought fiercely when they met. After a terrible battle, the Brown Bull of Cooley eventually killed the Finnbhennach but was himself mortally wounded. He wandered the land with the shredded carcass of his enemy hanging from his horns, dropping pieces wherever he went.

At Athlone the Brown Bull stopped to drink at a ford, and the loin of the Finnbhennach fell to the ground. The place from then on became known as Áth Lúain – ford of the loin. Other places in Ireland also took their names from the wandering of the Brown Bull and the fallen body parts of the Finnbhennach: Finnleithe, where his shoulder was thrown; Troma, where his liver fell; Étan Tairb, where the Brown Bull rested his head against a hill; Gort mBúraig, where he pawed the ground; and Druim Tairb where he finally died.

Sadly many of those names were changed in later centuries through the Christianisation of Ireland and the Anglicisation of place-names.

33

Leitrim

The wind was blowing fiercely, the trees bent almost double before its unrelenting gusts, and every now and then the skies unleashed fierce torrents of rain. This was the tail end of a hurricane that had whipped in off the Atlantic overnight and was raging across the land. Trees and power lines lay fallen in its wake and a caravan in Galway had been swept off a cliff, tragically killing its occupant. It is safe to say these were not ideal camping conditions, but camped we were.

The evening before, we'd flagged down a pick-up truck on the road to Drumshanbo. The man driving it had offered us a bit of wasteland by the River Shannon for the night. It was a little off the beaten track, but it was a pleasant, sheltered spot, he promised.

We followed his directions, making our way along a rough road through a forestry plantation for half a mile before we eventually fetched up beside the river. There was a large fenced-off area of hard standing and on either side of it were empty fields. We turned Micheál and the horses away in one of the fields, pitched our tent on the hard standing beside the trees, and set about making a fire on which to cook our dinner.

The sun shone, the world was still, and I wandered happily along the banks of the Shannon, looking out across the boggy pastures where little thorn trees grew amongst the yellow-brown of dying grasses. The flat plains of Roscommon had given way to rolling hills that climbed steadily towards distant mountains, which lay half-hidden behind a curtain of falling rain. A rainbow spanned the dark sky in the north, bridging the gentle river, and a pair of wild ducks drifted idly through the reeds on the far riverbank.

In the early hours of the morning the wind picked up and the rain pelted. The tent billowed wildly, straining at its guy ropes, and I wondered doubtfully whether camping beside a forest had really been such a good idea after all; but luckily no trees fell on us as we slept.

When we got up in the morning the wind was still blowing strong. Gerry Gilhooly – the man who had offered us this pleasant camping spot – arrived with his lovely wife, Mary, bearing a pot of tea, flasks of hot water, and bread and jam for our breakfast. They suggested we stay put for the day because the wind was stronger out of the valley, and the roads were in chaos. Not needing to be told twice, we all had another day off.

Mary kindly invited us to sleep in their caravan that night and said if the electricity hadn't been out, we'd have been welcome to a shower. We appreciated the kind offer, but chose to stay where we were. It was a beautiful spot and we were well settled.

By mid-afternoon the winds died down, the rain vanished, the day improved, and by the next morning we were ready to resume our journey.

We passed many fallen trees and scattered branches as we headed into Drumshanbo where we stopped at the shop to restock our supplies and to chat with some of the friendly locals. Then we left the town behind us, following a small road which led up into the hills above Lough Allen.

A few miles away to the south – too far off the route to warrant a detour – lay two hills made famous by the 17th century blind harper Turlough O'Carolan. They were the Fairy Hills Sí Bheag and Sí Mhór – the small fairy hill and the big fairy hill. Today, the tune is played the world over. Some say that O'Carolan once fell asleep in a fairy rath and from that day forward the fairy music ran in his head. Certainly they are some of Ireland's most beautiful melodies.

Not so very long ago, words were put to the tune of Sí Bheag Sí Mhor and a story was woven about those hills. The lyrics tell how beneath the hills lived two fairy clans who were constantly at war with one another. Although they fought and many were slain, neither clan ever won the battle. One day a prophet came to the hills, and he said to the clans that they should put an end to this pointless warring amongst themselves and unite to fight the common foe.

I first heard the song sung by a musician from Donegal. He told me that this seemingly innocent ballad about the fairy folk was actually a song calling for the people of Ireland to stop fighting one another and instead to join forces against their common enemy: the British.

We slept in a hay barn that night and in the morning our pleasant meandering through the scenic hills came to an abrupt end, when the lane we were following brought us out onto the main road beside the lough. We followed it for several miles and eventually arrived in Dowra, where we stopped for lunch – which we shared with all the animals. Apples and fresh bread rolls were enjoyed by all, and crisps, too. The horses both turned their noses up at the strong salt and vinegar flavour but, to our amusement, Micheál thoroughly enjoyed them and became positively friendly whenever he heard the suggestive rustle of a crisp packet. Ginger biscuits, we discovered, were also a firm favourite of our little one-eyed Irishman.

It was market day so the town was full of Land Rovers and pick-up trucks pulling trailer loads of sheep and cattle. All the pubs were heaving, even though it was barely lunchtime. The local policeman, keeping a watchful eye on things, came over to ask what we were about and a pleasant chat was had; then an English woman appeared who said she had travelled from Dublin to the west of Ireland in a horse-drawn wagon back in the seventies and had never left. This seemed to have been rather a common occurrence back then, because she was at least the third person we'd met to have done that.

The road from Dowra led us up into almost deserted hills. The vivid green of lowland pastures gave way to rough and boggy uplands; grey-green, windswept hillsides were mottled with brown patches of dying heather and bracken; and here and there the dark green of Sitka spruce and Douglas fir plantations stood out on the rolling landscape.

The views were spectacular, the sun shone, and we drank in the scenery in happy silence.

Coming down from the hills towards Glenfarne, we met a lovely couple – Joy and John – who let us camp in their barn that night and even offered us a shower!

Joy, who was originally from Cornwall, was animal mad. She had at least four dogs, and her yard was filled with an assortment ducks, chickens and geese. She also had six donkeys.

In Ireland, she said, if you had two donkeys in a field, it wouldn't be long before they became three or four as people dump their unwanted animals off in passing. Vlad's ears pricked and I could see him thinking how nice it would be to do the same with Micheál. I gave Vlad a reproachful look, followed by a 'Don't you dare!'

Micheál had actually been quite well-behaved of late – as well-behaved as could be expected, anyway. At long last he seemed to be adapting to life on the road, and a routine had been established. The day would start with us catching the two horses and taking them out of the field. Then, while Vlad groomed them and began the lengthy process of tacking up, I would spend a good half hour trying to catch Micheál. I'd offer him little bribes of apple and carrot, which he'd willingly accept, but then hastily retreat before I could catch hold of his lead rope. I'd taken to leaving a rope on him every evening when I turned him away to make catching him easier in the morning. Even so, getting hold of the mule was no mean feat. He had soon learnt to manoeuvre the rope in such a way that I couldn't just walk up and grab it without his agreeing to it first.

When the supply of treats ran out, Micheál would start trotting small circles around me – always just out of reach yet never trying to run away. Sometimes he'd stop running and let me get almost close enough to catch him before setting off at a spanking trot to circle me once more. Depending on his mood, we could have several such false starts in the morning. Then, when he finally tired of this game, he'd make a bee-line for the fence and pause for a moment, before pushing his way under it (whether it was electrified or not!) and go to stand quietly by the horses, waiting to be caught, groomed, and loaded up with his packs for the day.

Although this ritual was a daily test of my patience and perseverance, I suspected that the exercise was Micheál's none-too-subtle way of reminding me that he was here only because he chose to be. The terms and conditions upon which we had the pleasure of his company were negotiated by him and him alone, and thus any big-headed illusions I may have had about being in charge of the little creature soon withered and vanished. Never again, however, did we spend three hours trying to catch him, and for that I was most grateful.

From Glenfarne we made our way over the boggy hills to Kiltyclogher. Houses here were few and far between, the roads were quieter, and the landscape felt rural and uncluttered.

We passed several signs for Seán Mac Diarmada's house on the road, and when we arrived in Kiltyclogher we found a monument listing those who had fallen in Leitrim since 1916 under the Mac Diarmada's famous quote: 'I die that Ireland may live.'

Seán Mac Diamada had been born in Leitrim, not far from Kiltyclogher. He'd been an active member of the Irish Republican Brotherhood and had led the Dublin Easter Rising in 1916. He was one of the seven signatories of the 1916 Proclamation, which declared an Irish Republic.

Seán Mac Diarmada was executed for treason by the British Military on the twelfth of May, 1916 – along with many others who had been involved in the Easter Rising. The events of 1916 set the wheels in motion and three years later, in 1919, the Irish War of Independence began. To this day Seán Mac Diarmada is seen as a national hero and a key figure in Ireland's fight for independence.

34

An Afternoon with the Lennons

'It's a long struggle,' the old man said wearily, looking over the violin. 'This instrument doesn't want music to come out of it; it defies you to take out the music.'

We were in Rossinver at the house of Ben Lennon, one of Ireland's finest fiddle players. The shopkeeper in Kiltyclogher had suggested we pay him a visit and, as it was only a few miles off our route, we decided it would be well worth the detour.

Although he was ninety years of age, Ben was still playing his fiddle in the pubs and local sessions most weekends. He'd been born in Kiltyclogher and came from a family of musicians: his mother had played piano, a bit of accordion, and even some banjo, while his father and uncle had both played the fiddle. Following in his father's footsteps, Ben had taken up the fiddle at the age of ten, and – like his equally renowned brother Charlie Lennon – had gone on to enjoy a successful musical career.

Now, sitting in the front room of his house beside a roaring fire, he was happy to wax lyrical and share some of the wisdom he'd accumulated in his eight decades of playing the violin. He had a thick Leitrim accent, a deep voice, and a lovely turn of phrase. When he spoke it was slow and measured, almost as though he were speaking to himself. I clicked on my little voice-recording device and sank back into the deep sofa, listening intently to what he had to say.

'The Irish music is complicated,' he began slowly. 'Especially reels are not easily played.' He paused for a while. 'There's a big difference, now, between the classical and the traditional player,' he continued, thoughtfully. 'The classical trained person doesn't really have – or has difficulty in having – what the traditional player that worked by ear has developed. They just read the music and they can't go outside those parameters, whereas the traditional player has the freedom to do all kinds of ornamentation.'

I nodded in full agreement, hanging on his every word, and watching the flames dance in the fireplace. The room was stiflingly hot.

'There used to be many different styles – what we called local styles; but that's changed,' he said remorsefully. 'It's all become very global now. You have Chinese people playing Irish music, Japanese people – we've lost the regional styles.' He shook his head and sighed. 'I think to be a good player it has to be in your blood!' He stressed the word 'blood' vehemently, with passion, and a heavy silence followed. 'So, take out your violin now and let me have a look at your fiddle,' he said, suddenly brightening and eyeing up my violin case.

I did as I was told and handed him the instrument. He turned it carefully in his old, gnarled hands – scrutinising every inch of it.

'The arching is high,' he said disapprovingly. 'And the sound post is in the wrong place.'

He frowned.

'That's very important. You've got to get that seen to,' he said, sternly. I felt myself go red, mortified that I had presented this famous musician with such a poor specimen of a fiddle.

The violin had been bought for twenty pounds in a junk shop by the guitarist of my old band. It certainly wasn't my best instrument, and there was plenty of room for improvement – but it was fine for chucking on the back of a horse and carting around Ireland for hundreds of miles in all weathers! Even so, I suddenly regretted not bringing my other, better violin along with me.

'Not a bad looking instrument,' he concluded as he finished his examination. 'But this sound post…' he shook his head, then handed the instrument back to me and asked me to play him something.

Nervously I launched into the first tune that came into my head – Morrison's Jig. Ben listened attentively.

'She's a very good player,' he said to Vlad, sounding rather surprised. 'She has good control, but I'd not be playing it that fast, now.'

Then he picked up his own fiddle and showed me how he would play the same tune in the local style, using slurs and carefully placed rolls to give the melody a more lilting feel than my rather more detaché bowing. Hesitantly I began to copy his style, adding in slurs and five note rolls to ornament and accentuate some of the longer notes.

When we'd finished Ben nodded his approval.

'Give me your fiddle – I'd like to play a tune on it,' he said after a while.

He played us some airs, hornpipes, jigs, and reels – his fingers deftly running over the strings, as well as ever they'd done in his youth. He told me about each tune too – of the people who had written them, and who they were named after, and why. He had a wealth of knowledge, and I soaked up everything he had to say as the beautiful music echoed around the little house.

'There'd be nothing the matter with this instrument if it were properly set. It's a very sweet fiddle,' he conceded, passing the violin back when at last he'd finished playing.

At that moment a tall, slim man with grey hair quietly entered the room.

'This is my son, Maurice,' Ben said, introducing the man.

Maurice Lennon was also a well-known fiddle player. He'd co-founded the band Stockton's Wing, in which he'd played for many years, touring all over the world. Now he lived in the States where he taught violin and wrote music, but he was over staying with Ben for a while.

As well as being an excellent player and composer, Maurice knew a fair bit about fiddles, too, and after examining my instrument, he said he might be able to enhance the tone.

'I'll just need five minutes with it. I promise you I won't destroy it,' he said, and he vanished from the room.

While we waited for Maurice to return, Ben picked up his fiddle again and played me some more tunes – local ones this time: John McGovern's Reel, named after John Francis McGovern who was a stonemason from Kiltyclogher, followed by some tunes he himself had composed.

'Have you written many?' I asked him, but he shook his head gravely.

'My brother, Charlie Lennon, he's a terrific writer, but I'm not in that league at all. I'm not a swimmer and I don't go to the deep end of the pool – I'm paddling at the shallow end.'

After some while, Maurice re-entered the room and handed me my fiddle.

Hesitantly I drew the bow across the strings. It seemed louder, the tone was richer and it filled the space.

'What did you do?' I asked in amazement, unable to see any obvious difference in the instrument.

'It's something that was given to me by a luthier in the States and I'm not allowed to reveal what it is,' he said, mysteriously – giving me a long, intense look. 'It's tricky and I only do it occasionally, but if the moment feels right I'll go for it. What I can say is that within violins frequencies get trapped and I'm able to release them but I can't tell you how. You'll enjoy playing it a little bit more because it'll respond to you better now.'

He wasn't wrong. The music flowed out of the violin – loud, clear and resonating. It was beautiful. What had he done to make such a difference?

Maurice then joined Ben and together they played me some tunes that Maurice had composed: a hornpipe he'd written for Ben, and another – Master Shanley's – that Maurice had named after an old schoolmaster in Kiltyclogher who had encouraged him in his music.

When they finished, we all played some tunes together.

'If you had her for a month you'd have her up at the top,' Ben said quietly to Maurice.

'She's very close,' Maurice agreed, and I felt myself blush red, wondering whether they were just trying to be nice. Self-doubt is a horrible thing.

'If you leave me your email address I can send you some tips for improving your technique,' Maurice offered kindly. 'That's why you're here, perhaps. Everything happens for a reason. Don't be afraid of that. Embrace it. You've got a lovely bow-hand, and your tone is just perfect. Your intonation is really spot-on.'

'She's top class,' Ben said again.

'Absolutely,' Maurice agreed and then added: 'He doesn't say that lightly, you know. You have definitely got something special.'

I walked back to camp on a cloud, with the music and high praises of those two master fiddlers ringing in my ears. My enthusiasm for a musical career I'd given up on was renewed and at an all-time high.

35

Borders

It was past midday when we left our stop at a farm near Rossinver, where Declan Sweeney had offered us a place to camp.

We followed the main road towards Garrison, which skirted the shores of beautiful Lough Melvin. Halfway along it we came to a little bridge spanning a small river and suddenly we found ourselves back in the UK in County Fermanagh, Northern Ireland. Speed limits went from kilometres per hour to miles per hour, road signs were printed in a more familiar style, and our mobile phones reverted to UK networks. Other than that, not a lot changed.

By the time we reached Lower Lough Erne we'd covered fourteen miles and it was getting late. Time to stop, we decided, as we picked our way along quiet lanes that ran parallel to a busy main road on the northern shore of the lough. We started looking around for someone to ask, but most of the houses we'd passed had electric gates at the bottom of long drives and we didn't see a soul, so we traipsed on.

Eventually we met a jogger and I flagged him down to ask about somewhere to camp. He was a cheerful, enthusiastic man, with a heavy East London accent, which caught me totally off guard. Without a second thought he invited us back to his place. There was a shed we could sleep in and there were six acres of fields. They weren't actually his fields, they were his landlord's; and there was a herd of cows in them at the moment, too – but he felt sure it would be fine!

He sounded bright and optimistic, but I wasn't so convinced that his landlord would be agreeable to two random strangers putting some horses and a mule in with his cattle. Nevertheless, we took directions to his house and continued on.

Less than half a mile further along the road we passed a man outside a house and stopped to ask him about grazing for the night. When he said we could use the empty field opposite his house, we accepted. This seemed a safer bet than the jogger's offer, so we unloaded our gear and turned the animals loose. Although on a slope the field was waterlogged and we found ourselves standing in several inches of muddy water. In the middle of the field stood a small stone barn that was falling into disrepair. Feeling hopeful, we splashed our way over to have a look, but it was full of clutter and bird droppings and there was no room to move, let alone to pitch a tent or even spread our sleeping mats. Slowly, we began the hopeless task of trying to shift some of the clutter to make enough room to camp, but it was a futile undertaking.

We were just resigning ourselves to the miserable idea of pitching our tent in two inches of freezing cold water when a vehicle pulled up in the lane below the field. Out leapt our cockney jogger friend, Paul. He insisted we come back to his place, we were welcome to sleep in a shed, and his landlord had said it was ok to put the horses in the pasture – just so long as we shut the cows in another field. Without hesitating, we slung all our gear gratefully into the back of the van, caught the horses and Micheál – who were just as keen as we were to get out of the mire – and we set off down the road in the gathering dusk.

Paul lived with his lovely wife, Catherine, and their two sons, Thomas and Michael, in a large house set well back from the road and surrounded by several well-drained fields. The horses and mule were much happier with their grazing arrangements here, and Vlad, Spirit, and I were more than happy with the nice dry shed we were offered. It was certainly better than pitching the tent in a puddle!

Fiddler on the Hoof

When we'd set up our sleeping mats on a clean patch of concrete floor, we made our way down to the house where showers were offered, and a hot meal to boot! Even the washing machine was put at our disposal.

We basked in this unexpected kindness and luxury, and a pleasant evening followed as we got to know our hosts over dinner.

Paul was a painter and decorator, and to our surprise, we found that he could speak Romanian. He'd volunteered in Romanian orphanages back in the 90s and had picked up a lot of the language. While he and Vlad chatted happily away in Vlad's native tongue, I talked with Catherine.

Catherine was a talented mixed media artist, and some of her stunning pieces adorned the walls. They were vibrant works of art with rich, contrasting colours and textures. Each piece told a story or symbolised something deep and meaningful. Catherine herself was a quiet person, with a nice, calm way about her, and it was a strange contrast to Paul's boyish enthusiasm.

I jokingly asked her if she minded Paul appearing out of the blue with two weary travellers and a small herd of animals in tow. Catherine just laughed and said she was used to it by now. Apparently Paul was forever picking up waifs and strays – both human and animal – and bringing them home with him. He had a soft heart – it was what she loved about him the most.

The boys, Thomas and Michael, were both very wary of us and hovered about shyly in the background as we ate, drank, and enjoyed the good company of our kind hosts. Catherine said it was because they were forever being warned about 'Stranger Danger' at school and both thought their parents mad for inviting us in. While our hosts thought it sensible to warn children of the dangers that some people might pose, they also felt it was good to be able to show the boys that kindness and hospitality are important values to have, too, and that not every stranger is a threat.

The shed was bitterly cold that night and Vlad slept badly in his thin, summer sleeping bag. He hadn't wanted to bring a thicker one with him when we left Cornwall because the weather had been so hot, but that was a decision he was starting to regret. With these cold nights, we were very grateful for the woollen blankets that we used under the saddles because now they could double up as extra bedding for us. We would spread one blanket under the sleeping mats for extra insulation, and the other we'd use to cover ourselves in the sleeping bags. They worked a treat and at this point in the journey we no longer minded the fact that they were stiff with dirt and horsehair, and stank of months of sweat.

In the morning we enjoyed a hearty breakfast of porridge and fruit with our hosts before packing up and hitting the road again.

We were sorry to say goodbye to Paul and Catherine. They were such warm, caring people and we felt blessed to have met them – albeit for such a brief period of time.

Six times that day we crossed the invisible line that separates the Republic of Ireland from Northern Ireland and the United Kingdom. All that marked the border was a sign, and still the only differences we saw were in the speed limits and the road markings.

Not so very long ago, such unobstructed crossings would have been virtually impossible and many of the roads we followed that day would have been closed. Along the whole of the Northern Irish border – which stretched for more than three hundred miles – there had been a mere fifteen approved customs posts and checkpoints and most of the smaller roads crossing the border had been closed off, or even blown up, to prevent the illegal transportation of goods. Smuggling had been commonplace back then as many things were cheaper on one side of the border than on the other. Some of the older people we spoke to in the borderlands told how the women used to smuggle butter, tea, and washing powder in their undergarments, but not all the memories were so light-hearted.

In 1993 when the European Union launched the single market to allow free movement of goods, services, people, and capital, the Northern Irish border became redundant. The customs posts and checkpoints were gradually removed, and all the small roads were reopened. With the UK's imminent exit from the European Union, however, the future of the border was now uncertain.

Everywhere we'd been in Ireland people had asked us about Brexit, what we thought of it, and how we felt. It was a topic we hadn't dared to discuss in England and Wales where opinions were strongly divided and fiercely defended on both sides. In Ireland, however, opinions were unanimous: the decision to leave the EU was downright ludicrous. Both Vlad and I thoroughly agreed.

'People make fun of the Irish and say we're all thick,' one man in a pub in County Kerry had told us when the inevitable topic had been raised. 'But the Irish people would never have been so stupid as to vote themselves out of the EU!'

While Brexit had been a point of interest and a potential inconvenience to the Irish further south – mostly because it might create a trade barrier between Ireland and mainland Europe – in the north, it was a cause for great concern.

The dismantling of the border and the removal of checkpoints had been instrumental in establishing peace in Northern Ireland not so very long ago. People here remembered the Troubles all too well – the brutal violence and bloodshed that had only recently been put to rest. Everyone we spoke to here was worried that the hard-won, fragile peace would be lost, and the fighting would start up all over again. The situation, they felt, was volatile; already there were restless stirrings. Nobody wanted a border reinstated here, and most people in Northern Ireland didn't want to be removed from the European Union, either. Under the European flag the mentality of 'us' and 'them', Irish versus Brits, Republicans versus Unionists, was somehow less potent; the divide between the Republic of Ireland and Northern Ireland was diminished because both were part of the bigger picture, the greater whole. The borders were gone, and the physical signs of division had vanished.

All of that now hung precariously in the balance, ready to disintegrate at any given moment, and the people here were angry and resentful. They felt that their voices were being ignored in faraway Westminster, and their concerns swept under the carpet while ministers bickered amongst themselves over trivialities.

36

Donegal

That night we found ourselves in Donegal in the Republic of Ireland. We'd started looking for somewhere to stop several miles north of Pettigo but most people we spoke to said they couldn't help. Finally we flagged down a passing van just as we emerged from a patch of forestry next to a small, single storey cottage. We had the usual conversation with the driver about where we were going and what we were doing, during which he sussed us out and considered offering us a place to camp.

While we were chatting, the figure of a tall, squarely built man appeared around the side of the house. He was red-faced and rather unsteady on his feet. As he came towards us up the drive he stumbled, almost falling face first into the gravel.

'Why don't you ask Francie here if you can camp,' the van driver suggested brightly. 'I'm sure he'll help ye's out. Isn't that right, Francie?' he called out to the man, who was now trying to carefully balance crusts of stale bread on a bird table.

'Ah sure, work away,' the man slurred. 'Come in when you're ready.' And off he staggered.

I looked at Vlad and shrugged. We might as well. There was plenty of grass in the overgrown garden to keep the horses happy for a night.

Fiddler on the Hoof

We blocked off the open gateway with electric fencing, unloaded the horses, and then went round to the back door of the house. It opened straight into the kitchen-cum-living room.

Francie was sitting in an armchair in front of the blaring television, and on the sofa opposite sat a small, thin woman with long, straggly black hair. Both were drinking large glasses of rum and coke.

'This is Wendy,' Francie said gesturing towards the woman, who offered us a drink and told us to help ourselves to biscuits and toast. That was all they had in at the moment.

We settled in beside the turf stove, drinking cups of tea and getting to know our hosts. It was Wendy's birthday, and they were celebrating – just the two of them, they said. If the heaps of bottles and cans lying around outside the back door were anything to go by, the celebrations had been on-going.

I don't believe in taking advantage of people, especially not those who are in what can only be described as a compromised condition, but I do occasionally see the benefit of trying my luck. We hadn't slept in a house since leaving Mountshannon over two weeks earlier, I was tired of sleeping out in the cold and of setting up and packing down a tent, so try my luck I did.

Our hosts were well past the point of caring, and with some carefully steered conversation we managed to secure the offer of a bed for the night in the spare room – which was all kitted out for a little girl, with pink walls and Disney cartoon bed clothes. I didn't like to ask where the occupant was, because clearly she didn't live here.

We spent the evening drinking and chatting with our hosts, who were quite pleasant drunks, and after a while I played them some tunes on the fiddle which went down a storm. At about ten o'clock Francie staggered away to bed. I happened to be in our bedroom, searching through the packs for my journal in which to write up the day's events, when he tottered past. As I was heading back towards the kitchen, he called out from his dark bedroom:

'Is that you, Cathleen? Come here to me now.'

'What do you want?' I asked, sticking my head cautiously round the door.

Donegal

'Come here and give me a cuddle,' he said, with arms outstretched from amongst the crumpled bedclothes.

Somewhat taken aback, I put on my best no-nonsense attitude – which months of dealing with unruly elderly gentlemen in my care job had perfected – and told him firmly that we would be having none of that sort of behaviour, thank you. The next time Francie called out I sent Vlad in to him, and Vlad – who is much more patient and forgiving than I am – obligingly gave the man a hug, and tucked him into bed. Satisfied, our host fell asleep.

While Vlad was out of the room tending to Francie, Wendy suddenly leant over and grabbed hold of my wrist with a strength that caught me by surprise.

'Is he good to you? Does he treat you nice?' she asked, staring into my eyes with a fierce concern. I assured her that Vlad treated me just fine and she let go of my wrist, settling back into the sofa again with a satisfied sigh.

She was a timid little woman and there was something quite childlike about her; she seemed somehow ... broken. I wondered what sort of life she'd had, and whether perhaps it was the drink that had made her like this, or whether it was life that had led her to drink. Perhaps it was a bit of both – not that it was my place to speculate or judge! I also wondered, as we turned in for the night, whether our hosts would forget that they had guests when they sobered up in the morning. I needn't have worried, however, for by the time we got up, both Francie and Wendy were already in the kitchen watching the Jeremy Kyle show and drinking beer to cure their sore heads.

Vlad and I declined their offers of alcohol to start the day, being more inclined towards cups of tea and coffee first thing in the morning. After breakfast we loaded up the horses, thanked our hosts, and wished them well before setting off once more.

The landscape grew wilder and bleaker as we made our way northwards. Gentle hills became sweeping moors which climbed towards barren mountains. The wind blew constantly, and rain was frequent and heavy. One night we made camp in an empty sheep barn where the wind rattled the enormous metal doors all night long; another night we slept in the well-insulated but mouse-infested tack room of an empty stable barn; and once even, near Raphoe, a kind family gave us a room in their B&B, and offered us a shower and the use of their washing machine. Thus we managed to stay warm and dry in spite of the cold, wet weather. Both the barns and the B&B were considered great luxuries, because neither were expected nor guaranteed, and for the hundredth time on the journey we were humbled by the kindness and hospitality of total strangers.

The hillfort of Grianán Aileach marked the entrance to the Inishowen Peninsula, at the top of which – a mere thirty-five miles away – lay Malin Head: the northernmost point in Ireland and the end of our journey.

Grianán Aileach is an enormous ring fort which stands high on Greenan Mountain, overlooking the surrounding countryside for many miles. Legends say that the fort was built by the Tuatha Dé Dannan, when the Dagda – the god-king – had ordered it to be raised over the grave of his murdered son, Aodh. Other legends say it was a place of worship for Graine, Celtic goddess of the sun, from whom it takes its name.

We climbed a long, steep pass over some mountains and then dropped down towards the town of Buncrana. The sun shone here and there and the day was dry, but the cold wind was unrelenting. Before us and behind us lay the grey sea and the wild, heather-clad hills of Donegal stretched away on every side.

Skirting Buncrana, we followed a narrow lane which wound through low hills past small farms towards Dumfries and Dumfree. My legs were aching from the steep climb over the mountains earlier that day, so Vlad kindly offered to lead Micheál. After a mile or two, this arrangement was no longer agreeable to our little mule and he began objecting furiously. He hung back, dug his heels in, swished his tail, and swung wildly with his hind legs – but to no avail. His protestations were ignored. I was tired; it was nice to rest my weary limbs and ride Oisín and lead Dakota for a change. Anyway, it was high time Micheál and Vlad reconciled their differences.

Micheál, however, clearly felt otherwise and in a final, desperate act of defiance, the obstinate animal flung himself – shoulder first – into the front end of a car, which had kindly stopped to let us pass. Fortunately no damage was caused by this deliberate act of self-destructive vandalism. All hopes of the old hatchet being buried vanished on the spot so, sighing resignedly, I dismounted, and took up his rope again. Happy that things were all back to how they should be, Micheál tripped merrily along behind me, content.

That night we camped in the back garden of a man named James. The horses shared a field with a handful of young steers in a rough pasture below the house.

Once we'd set up the tent and all our gear was stashed safely away in the garage, James invited us in for a cup of tea.

The bungalow was a rather drab affair. It was dark and dingy inside and, although clean and tidy, was more functional than it was homely. Cups of tea were offered and fruit, toast, and bags of crisps were handed round. Then we all sat in the front room and chatted before our host headed off to Buncrana where he worked as a security guard at a nightclub.

James was a quiet, introverted man and conversation was an awkward affair with many long, uncomfortable silences. He'd been a farmer, he told us, but his wife had left him, taking all his land in the divorce settlement. He had a grown-up daughter, too, but she didn't speak to him anymore. He seemed lonely and a little sad, and he had a strange way of touching my shoulder, or taking hold of my hand and giving me long, searching looks, which made me feel quite uncomfortable. Perhaps I reminded him of his daughter, I thought.

Micheál was off-colour when we set off again in the morning. He plodded along, slowly and reluctantly, stopping often, but not wanting to eat. He carried his head low, and his ears – which were usually pricked and alert – were flopping out to the sides in a melancholy manner. I was worried about him. About a mile down the road we pulled over, and I removed Micheál's packs and fixed them onto Dakota's saddle so at least he didn't have to carry any extra weight. After that our mule brightened a little and picked up the pace. By the time we reached Carndonagh, he was back to his usual self.

Stopping off at a small supermarket on the outskirts of the town to stock up on supplies, we hitched the animals up to some convenient railings in a quiet corner of the car park and I stayed with them while Vlad went in to do the shopping.

It was here that we met Toni.

Toni Harkin had only popped down to use the public washing machines at the end of the car park that afternoon, but on seeing the unusual spectacle of two horses, a mule, and a wolf hitched to some railings, she had come over to find out what was going on. After a brief chat she invited us back to her house on the other side of town. Her sons were away at their father's that weekend, she said, so we'd be welcome to stay. Showers were offered, and a bed for the night, too; she even said we could put the horses in her garden, if they would fit.

The garden, we decided, would be too small for our three equines, but – undeterred, and determined to help us now – Toni set about finding a field for the horses nearby.

After a few phone calls and a little driving round, the horses and Micheál were eventually deposited in a field which belonged to a nice man named Derek. He had a couple of acres where a small flock of sheep were grazing alongside his coloured gypsy cob mare. The horses could stay for a night, Derek offered kindly, and our gear could go in the garage, too. It would be no problem at all.

Feeling grateful and relieved, we headed back to Toni's house.

Toni was a pretty woman in her mid-thirties and she lived on the edge of Carndonagh with her two teenage sons, Michael and Jack. She worked part-time in a call centre during the day, and at night she pulled pints in one of the many pubs in town. She had a wicked laugh, was animal mad, and although she had a heart of gold, I got the feeling that she had a sharp tongue and wasn't afraid to speak her mind. I like that in a person because you always know where you stand.

Showers were had, a nice meal eaten, and we slept soundly in a comfortable bed that night.

'Look, you guys,' Toni said brightly the following morning as we packed up our things and prepared to leave, 'the boys won't be home till tomorrow, you'd be welcome to stay another night if you like.'

It was a kind offer. It was only twelve miles to Malin Head, we could easily ride there and back in a day, and it would be nice not to have to lug our gear around for a change. We headed over to the field where the horses were and knocked on Derek's door to ask if that would be ok.

'Aye, sure, it's no bother!' he said cheerfully. 'Keep them here as long as you like.'

37

Finish Lines

It was the 30th of September. The day was bright and sunny but the wind was blowing hard, and every now and then heavy showers swept across the peninsula from the surrounding Atlantic Ocean. We rode through open countryside where little grew but wind-burnt grasses and dying heather. Bracken-clad hills rolled away in the distance and thick gorse grew beside stone walls that lined the fields bordering the narrow road. Trees were scarce, and the small handful that clung to a tenuous existence were dwarfed and bent from years of being battered by the strong winds.

To keep things simple, we'd left little Micheál behind with Derek's chunky coloured cob for company and had only taken Oisín and Dakota. I felt guilty leaving Micheál and wondered whether he'd think we'd abandoned him, but it would be faster and easier to ride the horses and would save my legs the long walk. Besides, it seemed fitting to cross the finish line on Dakota. This journey had been the making of him. He'd started out as a skittish bundle of nerves, terrified of the world around him, but over the many hard miles he'd grown into a calm, confident, sweet-natured little horse. It felt right that we end the journey together.

Micheál watched us from across the fields while we tacked up the horses in Derek's yard; as we set off down the road I heard his plaintive wail. My heart felt as thought it might break. He was an obstinate little creature, but I had grown to love him dearly.

Oisín's shoes – which had lasted over four hundred miles since he'd been shod near Castlemaine in County Kerry – were by this time razor thin and disintegrating rapidly. The nail-heads had worn out completely and one hind shoe clinked ominously as we rode, held in place by just two flimsy nails. I doubted it would hold for the twenty-four mile round trip we had ahead of us that day.

We took the main road out of Carndonagh and into Malin town, then followed little lanes which ran along the middle of the headland through bleak, wind-swept countryside for several miles. Beyond Ballygorman we picked up the coast road. The blue waves of the expansive Atlantic Ocean rolled, white-capped, under azure skies where huge clouds billowed. The wind whipped at our faces and on it we could taste the salty spray from breakers that crashed on rocky outcrops below us. The horses were keen and, although we trotted for several miles, neither seemed to tire. At last, ahead of us at the end of the long peninsula rose a hill on top of which stood a tall grey tower. This was Malin Head, the northernmost point in Ireland.

We raced up the hill to the top where, on the asphalt beside the signal station, a white line was painted with the words 'start' on one side, and 'finish' on the other.

We'd done it! We'd ridden the length of Ireland.

Toni had come to meet us with a friend. Cups of coffee were enjoyed and celebratory biscuits handed round, even to the animals who had travelled with us so faithfully all the way from Cornwall. Then, because the wind was sharp and cold, and the daylight limited, we turned round and headed back to Carndonagh, retracing our steps. When we got back, a less-than-impressed Micheál greeted us with a reproachful honk, and even Oisín – unable to help himself – let out a reassuring nicker. Secretly he was quite fond of the little beast, too.

Miraculously the loose shoe had stayed in place for the duration of the ride, but that was the last we saw of it. By the following morning it was gone.

We celebrated our achievements that night with Toni over a nice meal and a drink or two; then we fell happily into bed.

The next morning we awoke and realised that we had no plan. You might have gathered by now that planning is not our forte. Suddenly, after months on the road, we had no point on a map to aim for and no direction to take. Our only goal had been to reach Malin Head; beyond that we'd hardly begun to consider.

It was now October. Temperatures were falling rapidly, and the weather was becoming increasingly more miserable. After three months on the road we'd had enough of camping, and enough of traipsing about the countryside trying to find good grazing for the animals every night. I was starting to have fond visions of my cosy little caravan back home in Cornwall. And Taliesin – my faithful old horse! How I missed him! Yes, we decided, it was definitely time to go home.

With our sights set on Cornwall, the next step was to decide how to get there. We had two options:

We could hire a transporter to pick us up from Carndonagh – but after ringing several, we soon realised that not many were willing to come this far north. The few that were charged a lot of money.

We could ride south and try to find cheaper transport in the Irish midlands, a little closer to Dublin, but given the state of Oisín's shoes we would need to find a farrier first.

Before we could arrange anything at all, however, we needed to locate Micheál's passport because he couldn't leave Ireland without it.

We had sent off for it when he'd been micro-chipped in Mountshannon, and had decided to have the document sent to Mum's address. We could work out what to do with it once it arrived there.

Three weeks had passed and it still hadn't turned up, so a few days before we reached Donegal I'd rung the horse passport agency to see what was going on. Funnily enough it was at the top of their pile and was just about to be issued, they'd assured me. But by the 1st of October, the day after we'd reached Malin Head, there was still no sign of it.

We sat around the table in Toni's kitchen drinking our morning coffee and debating our options, trying to figure out how to proceed.

We were so caught up in our discussion that neither of us noticed Toni come into the room.

'Hey, look, if it helps ye's out of a fix,' she piped up suddenly from where she stood in the doorway, sporting a rather fetching dressing gown and fluffy pink slippers, 'I don't mind you stopping here for a few days until you get yourselves sorted.'

Vlad and I exchanged a hopeful glance. It would certainly buy us more time to sort out our next move.

Half-hearted murmurs of 'we couldn't possibly,' 'we don't want to be a burden on you,' and 'you've done enough for us already' were made, but our feeble protestations were rebutted, and Toni began to insist.

'Look, it's really no bother,' she said. 'Michael's away working this week so his bedroom's free anyway. Honestly, you're welcome!'

It was a kind and generous offer – although I fear it was one she may have gone on to regret over the days that followed. She was a good sort, Toni.

In the end, her invitation was gratefully accepted and planning and organisation began in earnest. Trying to locate Micheál's elusive passport, I rang the issuing body again, and they assured me the document was in the post, heading to County Clare as we spoke. That no longer seemed like such a good thing, so after a few phone-calls backwards and forwards and some rather frustrating bureaucracy, we finally managed to arrange for a duplicate passport to be issued to us at Toni's address in Carndonagh. It arrived first thing the following morning by special delivery. With that sorted, we now turned our attention to the problem of getting home.

Since neither of us wished to retrace our steps south, and nor did we see much point in putting shoes on Oisín for a mere hundred miles or so, we began ringing round transporters in search of one who was willing to collect us from Carndonagh.

Eventually we found a man who was happy to pick us up. He also said we could stay in the empty cottage on his yard until he had a lorry heading across to England. He could drop us anywhere along the M4 around Bristol, but from there we'd need to find our own way back to Cornwall. It was the best offer we'd had so far. Although the whole operation would set us back over £1,500 we decided it was worth it to get home, so we went ahead and booked it. He would pick us up the following day – Wednesday 3rd of October.

We awoke early and most of the day was spent getting packed and ready for departure. So far, everything was falling nicely into place.

In the late afternoon Toni dropped us off at the field with all our belongings to await the transporter. Fond farewells were made, and heartfelt thanks given for all her help and kindness.

We retrieved the rest of our gear from Derek's garage, piled it up at the side of the road, and then went to catch Micheál and the horses.

A short while later the transporter arrived. He was a thickset man with grey hair and small, piggy eyes, and a rough way about him. He surveyed our mountain of gear doubtfully, and then turned a disapproving eye on Spirit.

'You never said anything about a dog,' he said grumpily, and my heart sank.

I had definitely mentioned Spirit in our conversations over the phone – several times in fact – because I'd had to double-check with him that it would be OK to have her with us in the cottage. But somehow he must have forgotten it. This was not a good start.

Smoothing over that slight glitch, we began piling all our belongings into the vehicle. When that was done, we went to fetch the horses and began the process of loading.

Micheál went first, leaping agilely on board, and then Oisín followed willingly. Dakota, on the other hand, was having none of it.

He hung back on the rope, his eyes came out on stalks, and he darted about desperately trying to avoid treading the ramp. This upset Oisín who – concerned about his friend – began to object. He started pawing frantically and rearing, while whinnying loudly. All my efforts to calm him went ignored. Spirit, too, was kicking up a wild fuss, howling and yelping from her position amongst the loaded gear. It was pandemonium!

Suddenly, from the corner of my eye, I saw the transporter pick up a long stick from the roadside and before I could stop him, he had delivered Dakota a smart whack across the rump. Although it wasn't a hard blow, it was the worst possible thing to do to Dakota. He's a delicate soul who needs gentle, persuasive handling – not bullying! Panicked, Dakota shot backwards down the road, now flatly refusing to go anywhere near the ramp, or anywhere near the offending transporter, and Oisín's protestations became more fervent.

Then it was all too much.

'Right, that's it! I'm not doing this,' the transporter snapped, losing his patience. 'Get your horses off.'

I was stunned.

'I'm not taking them. Get them off before that one causes any damage,' he repeated, pointing at Oisín.

Powerless to argue, and still reeling from his harsh handling of Dakota, I quietly unloaded Oisín and Micheál and tied them to some nearby railings. All our gear was unloaded in silence and dumped unceremoniously at the side of the road. The transporter then demanded a hundred and fifty euros for his time and fuel.

In no mood for an argument, I glumly handed it over.

'Look, if you can make it to the yard, you're still welcome to stay in the cottage and I'll get you as far as Bristol,' he said, softening a little as he pocketed the wad of money. Then he jumped into his vehicle and disappeared off down the road.

We turned the horses out into the field again, lugged all our gear back to the garage, and then I burst into tears. What now?

'Don't worry,' Vlad said. 'I'll sort it out. I'll get us home!'

It was a nice sentiment; he meant well – but the words rang rather hollow.

'I don't care if he's the only transporter in Ireland, I'm not using him to get home! Not after the way he treated Dakota,' I said flatly to Vlad as we trudged wearily back into Carndonagh, and he agreed.

38

Conundrums

'I hope you're on your way back to mine!' Toni said as soon as she heard what had happened. 'I'm at work, I'll be back by eight. Get a fire going and make yourselves at home.'

Did I mention what an amazing and kind-hearted person she was?

After a subdued evening and a bad night's sleep, it was back to the drawing board to rethink our options. We quickly ruled out hiring another transporter to pick us up from Carndonagh. All the quotes we'd received from other transporters were nearly double the price of the last one, and what if we had a repeat performance with Dakota refusing to load and Oisín panicking and trying to destroy the lorry? We'd just end up losing more money we didn't have and getting absolutely nowhere. It was a risk we couldn't afford to take. No, we'd have to come up with another idea.

I will list just a few of the things we considered doing to get home and the conclusions we came to. There were many combinations of the following ideas, but it would be too much to include them all, so here is the gist of the main ones:

1) We could get Oisín shod and ride back down south.
Pros:
Finding transport would be easier and cheaper further south.

Cons:
We didn't particularly fancy another two or three weeks on the road. We were ready to go home.
Conclusions:
It was a viable option, but only as a very last resort.

2) We could hire a small self-drive lorry and move the horses and Micheál somewhere more central where finding transport would be easier and cheaper.
Pros:
We would get out of the inaccessible north and we'd have all the time we need to load Dakota in a nice, calm way with no bullying.
Cons:
Small lorries are only stalled for two horses so we'd have to do it in two trips.
It would be expensive.
What if Oisín tried to trash the lorry?
Self-drive horsebox hire was virtually unheard of in Ireland.
Conclusions:
It was an option, but not a good one.

3) We could buy a small three and a half tonne lorry and transport the animals home ourselves.
Pros:
We would end up with a lorry that we could sell to recover some of the costs.
If it was our lorry it didn't matter if Oisín destroyed it a little.
Cons:
Small lorries were very expensive to buy.
They're only stalled for two horses so what would happen to Micheál? (At this point Vlad said to leave him behind, but I wouldn't hear of it.)
Conclusions:
It was not a viable option (although Vlad disagreed).

4) We could buy a vehicle with a tow-hitch and a trailer and drive the horses home.
Pros:
We could sell the vehicles and recoup some of our costs.
Cons:
Neither of us could legally pull a trailer.
Trailers and cars were expensive and our budget was limited.
A trailer wouldn't be big enough for all three equines and I refused to leave Micheál behind.
Conclusions:
Not a viable option.

5) We could buy a seven and a half tonne lorry to get us home.
Pros:
All the horses could fit in it.
Seven and a half tonne lorries were cheaper to buy than three and a half tonne lorries.
We could sell it when we got home to recoup costs.
If it was our lorry it didn't matter if Oisín destroyed it and little.
Cons:
Neither of us could legally drive a seven and a half tonne vehicle, so we'd have to find someone who was willing to drive it back to Cornwall for us.
Conclusions:
It was a crazy idea.

The entire day was spent coming up with and discussing these ideas, evaluating them, and weighing up the many pros and cons and possible outcomes. All avenues were explored and all options were debated until – sooner or later – we came up against a supposedly insurmountable obstacle. Then the idea would be abandoned, and we'd move on to the next one. Thus it went until eventually we'd exhausted all our options and found ourselves back at the beginning, carefully re-considering each idea all over again and looking for a solution.

Conundrums

Round and round we went, tracing and retracing old ground and old ideas, re-thinking and re-evaluating all of our options, or coming up with new options – each one more far-fetched and ludicrous than the last, trying desperately to find a way to get home – but to no avail.

Frustrations grew, tensions rose, and heated arguments erupted. Sometimes they subsided, but often they descended into long sulks and resentful silences, broken only when one of us started re-examining the same old options again, looking for a solution that we might possibly have missed. The situation seemed utterly impossible.

Vlad was adamant that buying a seven and a half tonne lorry was our best option. He'd find someone to drive it, he promised. He had friends and family who could. Better yet, he exclaimed, he'd drive it himself – even though legally he couldn't because he didn't have a HGV license. Nevertheless, he was willing to take the risk. We'd drive to Mountshannon on the back roads where we wouldn't meet any Gardaí (the Irish police). Once there, we would have more time to find someone willing to drive us home.

Hell, Vlad said at one point, he'd drive us all the way back to Cornwall. It would be fine! What were the chances of being stopped, anyway? He could drive heavy construction machinery: diggers, rollers, dumpers, and forklifts. He taught people to drive those things for a living, and assessed them, too. What was a seven and a half tonne lorry? He was more than capable of doing it – he just didn't have the license, that was all. It was a mere triviality.

I was doubtful, but after a while the idea – although utterly insane – was beginning to look like a sensible one, because so far we had failed to come up with anything better.

We scoured the Internet and local selling pages for Donegal and Northern Ireland, looking at all the horse lorries that were for sale. There weren't many nearby. Most we found were much too expensive, had no MOT, or were in dire need of repair. We didn't have time to mess about with this stuff! We'd been here long enough. Poor Toni had only invited us to stay for a night, yet here we were five days later, still moping about in her house. Although she'd said we could stay as long as we needed to, I felt we were beginning to overstay our welcome.

Then at last I found something. There was a lorry for sale in Falcarragh, an hour's drive away on the next peninsula over. It was seven and a half tonnes, stalled for five horses, and at two thousand euros it was within our meagre budget. Although it was twenty-eight years old and looked in need of some T.L.C. and a good makeover, at least it was roadworthy. Its MOT was up to date, it had recently had new marine ply floors put in, and it had new tyres all round. It was a good little lorry, the lady selling it assured us, and it had never let her down.

I showed it to Vlad.

'No,' he said.

'Why not?' I asked.

'I don't like the look of it. It's old and covered in rust. Anyway, I can't drive it.' He shrugged, and went back to his phone where he was looking at trailers for sale and contemplating buying a tractor with which he could legally tow it.

Here was the man who, barely an hour before, had been willing to take a risk and illegally drive a lorry all the way back to Cornwall! But when push came to shove, faced with the reality of the situation, he suddenly backed off the idea.

'Well how else are we going to get home?' I snapped, frustrated. 'One minute you're telling me you'll drive a truck, and the next minute you don't want to!'

'Well what about this one?' he said a short while later, showing me a lorry on his phone that was twice the price of the one I'd found, and had no MOT.

'It's not even roadworthy,' I said, exasperated. 'Anyway, you just said you didn't want to get a lorry because you can't drive it!'

So round we went again, weighing up and discussing all the options ad infinitum. Arguments were put forward, counterarguments produced, and we came up against the same old brick walls until we were positively dizzy and our heads hurt. It was exhausting! Then things descended into another blazing row and finally Vlad had had enough. He disappeared upstairs and returned some while later with all his belongings stuffed into a bag. Without saying a word he made for the front door.

'Where are you going?' I asked, bristling.

'Home!' he shot back.

'What about the rest of us?' I asked, incredulous.

'You can find your own way back!' he said, opening the door and stepping outside.

'Well if that's the case, hadn't you better take your passport?' I snapped sarcastically. 'You won't get very far without it!'

Since Vlad had lost his driving license and credit cards all those months ago, I had taken charge of all the important documents. I knew with absolute certainty that his passport – along with mine, and those of all the animals – was tucked safely away in one of my saddlebags. He wouldn't have known where to look for it.

I half wondered in that moment whether it was wise to call his bluff and help him on his way. Maybe I should have let him go to see how far he got before he realised he'd forgotten his passport and came sheepishly back with his tail between his legs. But in that moment I was too livid and I don't do rational when I'm angry. If he wanted to go, then I certainly wasn't going to stop him!

Vlad stepped back inside, closed the door, and slumped down on the sofa at the far end of the kitchen, deflated. I knew he'd had no real intention of abandoning me and the animals to find our own way home. Vlad's not like that; he's too loyal, caring, and considerate – but I did wonder how far he'd have gone before he turned back.

'I'll take you to Falcarragh,' Toni offered kindly when she got back from work later that afternoon and I told her about the lorry. 'It's worth a look at least.'

In the end, Vlad was talked round so we bundled into Toni's car and set off.

Tina Reaney, the woman selling the lorry, met us and in the middle of Falcarragh and led the way to where the lorry was parked.

'This is Derek,' she said in a thick British Midlands accent, gesturing to where a lorry with grey metal sides and a blue cab flecked with rust stood parked on an area of hard standing. 'He might not look like much, but he's a good little lorry.'

'Why is it called Derek?' I asked.

'Well, he's like a council worker: he's slow and steady, but he always gets the job done,' she replied.

Her experience of council workers was rather different than mine. From what I'd seen, they spent most of their time parked up in lay-bys drinking tea, and on the rare occasion you witnessed them doing something, then there'd only ever be one man working, and about five others standing around watching.

The lorry wasn't actually hers, she told us. She was selling it for her old boss, who had recently closed down a riding school and sold everything that went with it, before heading off to the States. The ponies and tack had long gone; Derek was all the remained.

On closer inspection it didn't look all that bad. There was some rust here and there, the interior was a bit mouldy and needed some work, but it appeared sound enough and seemed like it was up to the job.

'Do you think it would get us to Cornwall?' I asked Tina.

'Definitely!' she replied, without hesitation. 'Jump in and we'll take it for a spin.'

A few hours later, after a lot of faffing about at cashpoints and drawing out as much money as our cards would allow, we finally managed to scrape together the asking price for the lorry. Receipts were drawn up and signed, keys and documents were handed over, and Toni headed home; then Vlad and I got into our new lorry, Derek, and began the sixty-mile drive back to Carndonagh, praying we didn't get stopped.

It struck me then that the whole situation was utterly ridiculous, and more than a little stupid – although this seemed to have been a running theme from the word go, so I had no reason to be surprised! Sane and sensible had long since been ruled out, and we'd had no choice but to take the plunge and hope for the best.

It was past midnight when we arrived back in town. Neither of us had eaten dinner, so we stopped at a takeaway for some chips and a vegan burger.

When Vlad went to fire Derek back up, the key turned in the ignition but nothing happened. The engine didn't even turn over. Shit.

Suddenly, taking a risk and investing all our meagre funds into a lorry didn't seem like quite such a brilliant idea anymore. We were two thousand euros down, had a lorry we couldn't legally drive, and the sodding thing wouldn't even start. What had we done? All the optimism I'd felt at having finally made a step towards getting home vanished and my heart sank into my shoes.

'Don't worry,' Vlad said brightly, ever the optimist. 'I'll sort it out!'

An hour later, after a lot of Googling and a phone call or two to some of Vlad's mechanically-minded friends, and the lorry still wouldn't start. This was just great!

We were parked on a slope and there was still enough air in the braking system so – taking a gamble and hoping it wouldn't wreck the lorry completely – we eventually rolled the lorry down the hill to bump start it. Luckily for us, the manoeuvre worked, Derek roared to life, and we made it back to Toni's where we tumbled gratefully into bed.

In the morning once again the lorry wouldn't start. Vlad – although somewhat lacking in both common sense and horse sense – is pretty good at practical things like mechanics. He knows a lot about engines, and there isn't much he can't fix. After his morning coffee he set to work locating the problem and soon discovered that the battery terminals were corroded. With a bit of tinkering and playing about with some tools the lorry was up and running.

Then it was back to the drawing board again to formulate a plan.

Now we had to figure out how to move forward from here. We had a lorry, so all that remained was to find someone to drive us from Carndonagh to Cornwall – or at the very least back to Mountshannon where we wouldn't be a burden on Toni and could work out the next move.

Vlad got on his phone and rang a few of his friends. Brent was working all weekend, Barry wouldn't pick up, it was too much to ask Phil, and Vlad's uncle was away driving lorries on another job. It was a huge favour to ask someone to drop everything and come all the way out here to drive us home.

With all the sensible options exhausted, the endless circular discussions resumed: fresh ideas were put forward and counter arguments produced. Most of our remaining options were downright crazy and not strictly legal, but with limited possibilities, we were now plumbing the depths in sheer desperation. Nerves were frayed, tempers flared, and by mid morning, once again, Vlad and I were barely on speaking terms.

I can tell you I've never had such an enduring headache, or experienced so much frustration in my entire life as I did in those few days in Carndonagh while we tried desperately to find a way home. It was Friday the fifth of October now and although it had only been a few days since we'd reached Malin Head and the end of our journey, it felt like an entire lifetime. The whole thing was proving more stressful than three months travelling on horseback had been, and more logistically challenging than riding a thousand miles from Cornwall to Ireland!

Suddenly, Vlad broke the heavy silence.

'I think I've found us a driver!' he exclaimed excitedly, showing me his phone.

One Adam Firks, member of the Alternative Living Group on Facebook, had posted an advert offering his services to all and sundry. He was at a loose end for a month and looking for work. Although based in the UK, he was happy to travel – any job considered!

'Erm, I highly doubt he'd be up for driving two random people and their horses from Donegal all the way to Cornwall,' I said, shamelessly raining on Vlad's parade. Then, seeing his crestfallen countenance, and not wishing to spark yet another argument, I quickly added: 'But it's worth asking. The worst he can say is no. Just make sure he can drive a seven and a half tonne lorry!'

39

Farewell to Ireland

At last we had a plan of action! It was almost too ridiculous to be true. Adam, the stranger from Facebook, had agreed to drive the lorry for us. We'd have to arrange his transport to Carndonagh, cover all his costs, and pay him a daily fee, but a more reasonable offer would have been hard to find.

Adam would be arriving first thing on Monday morning, so preparations for departure began in earnest: flights were booked, bus tickets bought, and the lorry was insured. We booked a ferry from Belfast to Stranraer in Scotland, then I arranged a place to overnight at Birchenhead Farm in Wardle, Greater Manchester, where I'd stayed with Taliesin on our journey home the year before. Next we needed buckets, water containers, hay nets, haylage, and feed for our equines.

There was a farmers' store just around the corner from Toni's so we picked up most of our supplies. Haynets, however, were not in stock, nor had anyone got some that we could have.

'Hey, what about old fishing nets?' Toni suggested. 'There's loads around the coast. Could you make a haynet out of one of those?'

It was a genius idea!

A quick trip to a nearby fishing village saw us return triumphant with some old nets and several lengths of good rope, which we'd salvaged out of a skip on the pier. Once we'd washed all the fish guts off of them and sewed the sides up with rope, they made excellent hay nets.

Water containers were expensive to buy, and money was now short, but Vlad had a bright idea. He walked down to the local take-away and asked for two empty oil drums. After a thorough wash, they were up to the job. It's quite amazing what can be achieved with a bit of resourcefulness, and it was nice to recycle items which would otherwise have been thrown away.

At long last we were sorted and ready to leave. Second time lucky!

On the evening before departure the tension was almost palpable and you could have cut the air with a knife. I had a tight knot in my stomach that twisted every time I thought about what lay ahead and I could barely swallow my dinner that night.

We'd placed all our trust in a complete stranger who might let us down at any given moment. And even if he didn't, there was so much else that could go wrong. What if the lorry didn't start? What if the horses refused to load? What if I couldn't catch the mule and we missed the boat? Everything hung by such a tenuous thread. I tried desperately not to think about it.

Looking for a distraction, I went with Toni to collect water from a nearby spring.

'Thank you so much for letting us stay, and for everything you've done for us,' I said for the millionth time as we sped along the narrow lanes in the dark. 'I'm sorry it's taken us so long to get sorted. You've been an absolute hero!'

'Hey, don't worry about it,' she said.

'I'm sure you'll be glad to get us out of your hair and have the house to yourself again!' I laughed.

'Look, it's been great meeting you guys, and it's been nice to be able to help. And actually,' she continued, 'in a way it's really helped me, too.'

'What do you mean?' I asked, confused.

'Well, sometimes I get a bit down and depressed,' she said. 'Helping you two has reminded me that I'm not a bad person.'

It had never occurred to me that kindness and hospitality wasn't just a one-way thing – that it wasn't just Vlad and I who were benefitting from it. Perhaps in this hectic modern age of self-absorption and ego-driven lifestyles, giving people the opportunity to take a chance on two strangers and invite them in from the cold and rain helped remind them of their humanity and the better side of their own nature. It was certainly an interesting thought.

Monday the 8th of October dawned, and I was awake early. I'd slept fitfully – that knot in my stomach was making me feel sick and I held by breath, half expecting everything to fall apart at the last minute.

Toni set off to fetch Adam from Derry, while Vlad and I saw to our last bits of packing and finished off the final touches in the lorry.

After an hour Toni returned with our driver in tow. Adam was a squarely built, middle-aged man with a round, cheerful face and slightly thinning hair. He looked pretty nice and normal – not the kind of person who leapt on a plane at a moment's notice to help two random strangers out of a fix, but heroes come in all shapes and sizes. Introductions were made, I handed him a coffee, then we piled into the lorry and headed over to the field.

Pre-empting some difficulty with loading, we'd allowed ourselves a good hour to get everyone on board and Toni had rallied round some of her equestrian friends to help.

We caught Oisín and Dakota without a glitch, but Micheál decided the whole thing was just too suspicious and wouldn't come near the gate. Even the seductive rustle of a packet of ginger nut biscuits did nothing to mitigate his distrustfulness. He trotted round and round the field, eyeing me up warily and swishing his tail in defiance any time I got too near.

In the end – running out of time and patience – we managed to corner him by the gate, one of Toni's friends grabbed hold of his rope, and at last we were ready to load up.

Oisín went first, following a bucket of feed. Once secured in his partition, he tucked happily into his haylage without kicking up a fuss.

Dakota, on the other hand, was predictably reluctant. A rattled bucket of feed could not inspire him to set foot on the ramp. He objected wildly, running backwards down the road, leaping about, and letting fly at anyone who happened to be standing too close.

Eventually, with a lot of patience, some lengths of rope, and a bit of gentle persuasion, we finally got him in. But we'd never have done it without the help of Toni's friends!

Last of all, Micheál leapt agilely up the ramp without protest to be with his friends. We were finally ready to go.

Hugs and heart-felt gratitude were given all round and Toni was thanked yet again for her patience, hospitality, and her unfaltering kindness and generosity. A more beautiful soul we could not have wished to meet! I sincerely hope that, if ever she finds herself feeling down, she remembers this: somewhere out there are at least two people who think the absolute world of her.

At last we were off and homeward bound, Derek trundling along at his top speed of fifty miles per hour, heading towards Belfast docks.

As I watched the scenery flashing past the window, I thought back over the miles we'd covered, the landscapes we'd seen, and the many wonderful people we'd met on our adventure. We'd come here to discover Ireland: seeking the myths and magic that had shaped this land and the people in it, inspiring generations of poets, writers, and musicians. And we had been successful in our endeavours! We had ridden the rocky headlands, followed the shores of quiet loughs, and crossed rolling hills which remembered the ancient gods and heroes of long ago times; we'd spoken to horse-drawn travellers around smouldering fires; seen the fairy raths and lone thorn trees in emerald green fields. We'd found music, too – jigs and reels, heart-felt ballads, and rousing songs of freedom. But more importantly still, we'd met the people of this land, been welcomed with open hearts and open minds, and experienced immense kindness and hospitality everywhere we went. The beauty and magic of this land would stay with us forever in the memories and friendships we'd made along the way, and the bonds we'd forged with our animals would remain unbroken in years to come.

What an adventure it had been!

Epilogue

We made it home – I should probably clear that up. Adam kept us entertained for the entire journey with fascinating stories of his adventures and his many moneymaking schemes. He was a sound person, with a wonderfully positive outlook on life and a real thirst for adventure. We are deeply indebted to him for taking a chance, hopping on a plane, and coming to drive us, along with our wolf, our two horses, and our one-eyed mule, all the way from Donegal to Cornwall. There aren't many people out there who would be willing to do such a thing!

We arrived back in Cornwall on Tuesday 11th of October at eight o'clock in the evening, where our weary equines staggered off the lorry in the dark. Two of them recognised the place they were in, and made their way happily into a familiar pasture full of grass. One little mule followed them warily.

Two days later, I went to fetch Taliesin and rode him the six short miles home to rejoin his friends and to meet the mule. He'd come sound by the time we reached Ireland – just a little too late to join the journey. He timed it all perfectly, but I'd long since forgiven him. Although he dropped us well and truly in it at the last moment, he gave Dakota the chance to have the journey of a lifetime. It facilitated Dakota's much needed education, rounding off all his edges and rebuilding my trust in him. It was a blessing in disguise. We'd left Cornwall with a semi-feral, dangerous lunatic of a horse, and returned with a calm, sensible road-horse – and a wilful, opinionated one-eyed mule for good measure!

In the aftermath of the adventure, Vlad was made an official member of the Long Riders' Guild – the organisation that had inspired all my rides and shown me that journeying on horseback wasn't merely a childish dream, but was actually a very real possibility. Vlad is the first Romanian to become a member.

As I write these final lines, I'm sitting looking out across the countryside. Acres of broom and lavender scrub with small clumps of pine dominate the landscape as it rolls away to hazy hills on the distant horizon. The sun is shining and it's a warm day – 17 degrees – although it's only February.

From where I'm sitting at a wooden table in a small room with large windows I can just make out Taliesin's enormous frame where he's dozing in the warm sunshine barely three hundred yards away. Oisín is there, too, hoovering up the rich spring grass like there's no tomorrow – his voracious appetite never satisfied; and every now and then Dakota wanders across my line of vision, swishing his tail against the pesky flies. His white coat looks grey – he's been rolling in dirt somewhere. And there – tiny next to the three horses – is a little black creature with white stockings on his hind legs, long ears, and one fierce little eye. Micheál still chooses to be with us, although he likes to regularly remind me that it is only upon his terms, and never mine.

Vlad has just put a steaming hot cup of tea down in front of me. Writing is thirsty work. He's still here, too – keeping me well-fed and happy. I finally stopped pushing him away and keeping him at arm's length. It took a thousand miles and then some before I finally accepted that he was here to stay, offering unwavering support to my wildest dreams.

I suppose when someone's willing to give up everything, jump on a horse and follow you into the unknown on an epic adventure, then surely they must be worth keeping, right? He's still not providing the level-headed voice of reason that I sometimes think I need – in fact some of his crazy ideas make mine seem quite grounded and reasonable – but perhaps that's a strength, not a weakness.

Across from the window where I'm writing, on a gentle rise in the ground, stands a seven and a half tonne lorry. Its grey metal sides and blue frame are rusting in places, and it could still do with some T.L.C.

and a good makeover, but Tina Reaney was right: he's a solid, reliable little lorry. He got us here, alright. I should probably add that Vlad can now legally drive it! And inside the lorry, dozing in her nice comfortable bed, is Spirit – enjoying her life of luxury for now.

It's February 2020; we're all in Portugal, waiting for the spring, and another epic adventure is about to begin – but that's a story for another time.

Additional Information

Total number of days on the road: 104 (3 months and a bit)

Total distance: 1023 miles/1646 kilometres

Number of days spent travelling: 64

Number of days off: 40

Shortest day: 4 miles/6 kilometres

Longest day: 25 miles/40 kilometres

Average distance per day: 16 miles/26 kilometres

Navigation: OS Landranger maps for England and Wales, mostly using the MapFinder app on my mobile phone. We had paper maps but it was too risky to have flapping paper on Dakota. Once in Ireland we used Google Maps.

Shoes:
- Dakota: 4 full sets
- Oisín 3 full sets

Training and fitness: regular rides for six to eight weeks prior to departure, increasing the daily mileage a little each week for Oisín. Dakota was completely unfit before we set off. He was walked in hand for much of the first few weeks. Ridden work was gradually increased throughout the journey.

Finding stops: As far as Stroud we were mostly staying with friends I'd made on previous journeys. A few extra stops were added in by putting posts on Facebook asking for somewhere to camp in the area we needed. In Wales we stopped and asked people, and sometimes we stayed with friends of previous hosts. In Ireland we relied mostly on the kindness of strangers and found almost everyone friendly and welcoming.

Equipment List

For the horses:
2 x saddles
2 x bridles
2 x rope head collars and lead-ropes
2 x pommel bags
2 x Trail Max saddle bags and cantle bags
2 x woollen blankets
2 x sheepskins
1 x thick western saddle pad (on Dakota)
Hi-viz leg bands
Hi-viz tail streamers
Hi-viz pack covers
Bungee cords
Hoof picks
2 x small, lightweight brushes
Basic first aid kit
Tick removal combs

For Spirit:
Hi-viz coat
Paw boots
Dog food
Poo bags

Corralling:
Collapsing fence stakes made from old tent poles

Electric fence wire
Insulators
Small battery-powered fencer (Hotline Shrike)

Camping:
Tent (Vaude 3 man tent)
2 x inflatable sleeping mats
2 x sleeping bags

Cooking:
Gas burner
Gas cartridges
2 x tin cups
Cutlery
1 x saucepan
Food (rice, pasta, couscous, oil, salt, pepper, tomato puree, spice mixes)

Clothing:
Several changes of travel clothes (trackies, t-shirts and hoodies)
Several changes of underwear
1 x set of nice clothes each for days off
Waterproof clothing
Hi viz hoodies and vests
Snoods

Electronics:
Mobile phones
2 x battery banks
1 x VHF Radio
Head-torches
Spare batteries for torches and fencer
Plugs and USB chargers for all devices

Repairs:
Sewing kit

Cable ties
Baling twine
Gorilla tape
Electrical tape
Superglue
Hot glue
Screwdriver
Multi-tool
Folding knives

First aid and hygiene:
Antibacterial wipes and hand gel
Plasters
Bandages
Knee-braces
Anti-septic creams
Sanitary towels
Nail knit
Kitchen towels
Soap
Deodorant
Toothpaste and toothbrushes
Hairbrush and hairbands
Small towel

In the pommel bags:
Snacks
Water bottles
Anti-bacterial hand gel and wipes
Kitchen towels
Gloves
Small plastic bags for litter and dog poo
Spring weight measure
Multi-tool

Repair kit containing: sewing kit, gorilla tape, cable ties, baling twine, Velcro, screwdriver, pair of scissors.
Head-torches
Vlad's VHF radio (licenced)

Miscellaneous:
1 x fiddle
1 x book
2 x diaries and pens
2 x digital voice recorders
Passports (human and animal)
Lighters
Rubber gloves
Woollen gloves
Spring weight measure for balancing packs

Also by this author

Before Winter Comes
A Journey on Horseback from Scotland to Cornwall

In the late summer of 2017 Cathleen Leonard set off on an epic adventure to realise a childhood dream. Taking her rescued draught horse, Taliesin, and her wolfdog, Spirit, she travelled over 1,000 miles from Durness in the northwest of Scotland back to her home in Cornwall. This is the story of one woman's journey of self-discovery, courage, determination, and encounters with the better side of human nature.

ISBN-10: 1070129100
ISBN-13: 978-1070129105

Available to buy from Amazon.

Printed in Poland
by Amazon Fulfillment
Poland Sp. z o.o., Wrocław